Climate Change in the Sunlight

by Rolf A. F. Witzsche

Contents

3

5

On the Ice Age and Climate Change
and the book

Climate Change in the Sunlight

Climate change is not visible in the sunlight, but the basis for it is. Climate change is forced by the Sun. The sunlight reveals that changes is solar activity reflect cosmic factors, external to the Sun, which the Sun merely responds to. The sunlight overturns the long-standing misperception that the Sun is a sphere of hydrogen gas that is powered from within by nuclear fusion of hydrogen into helium. It is physically impossible for such a Sun to produce the type of sunlight that we see. The phenomenon can only be created by a plasma-powered Sun, with plasma-fusion reactions occurring on its surface. Such a Sun is externally powered. Its actions fluctuate with changing conditions in interstellar space. The big climate changes on Earth are the result of that. Climate Change is never manmade. Anthropomorphic Climate Change is impossible. The climate changes on Earth are forced by the Sun via solar cosmic ray fluctuation that affects cloudiness. The solar effects on the Earth climate are even measurable in Carbon-14 isotope ratios. But the biggest climate change is yet to come, in the 2050s, when the plasma-focusing system breaks down under threshold conditions, and becomes inactive. In this inactive mode the solar activity reverts to the cosmic default level, its inactive state in which 70% less energy is being produced and the next Ice Age begins. The sunlight stands as an item of evidence for this extremely critical potential.

Numerous fields of evidence tell us that the next Ice Age is near. That's where the truth begins. Most of the evidence was discovered in the 1990s and thereafter. Some evidence is measured in ice cores; some is measured in space, by satellites. Some measurements are also made on the ground in terms of measurements of the Earth's magnetic-pole drift observed in northern Canada. All of this is seen combined with high-energy physics experiments at a leading national laboratory, and is also explored in the small in static experiments.

Against the background of these widely diverse types of evidence that have been recently discovered, the historic Little Ice Age in the 1600s, takes on a new dimension as a yardstick for measuring the future that by this evidence promises to be up to 40-times colder than the Little Ice Age had been. It qualifies for the term, Absolute! The evidence poses a great challenge ahead. Are we ready to respond? The Ice Age phase shift in climate is a stark in differences as night and day, and similarly fast.

In the Little Ice Age between 10% and up to 30% of the populations in Europe had perished by starvation. The last Big Ice Age was evidently vastly harsher. Only 1-10 million people emerged from it alive. That's all we had after 2 million years of development. We want to do far better this time around; and we can, with large-scale technological infrastructures for our food supply. But will we create them? Will we get the job done in the 30 years that we still have left before the Ice Age starts anew? Will we even consider it? And how certain are we that the phase shift to the next glaciation period will begin, as the evidence suggests, in the 2050s? We have no slack on this front. We have no slack on this front. Should we fail us on this absolute front, we would be committing suicide.

So, what will the answer be? Will we move with the evidence? Or will we lay ourselves down to die by default?

It takes an independent researcher to brake the taboos that have kept mainstream cosmology imprisoned, increasingly, during the past century, even while what is regarded as taboo is known to be wrong.

The Illustrated Science series is intended to open the scene beyond the threshold of accepted taboos, to where the actual physical evidence speaks for itself.

The scope of the existential challenge that the Ice Age brings with it, takes astrophysics out of the academic domain and places it into the foreground as one of the most-critical issues of our time. The big Climate Change events that have already worldwide effects are mere fringe effects in the flow of the ever-changing cosmic dynamics. The big effect, when the Ice Age begins anew, promises to be caused by a dimmer and colder Sun with 70% less radiated energy. This defines our climate future.

Sure, we can live with all that by creating new platforms for agriculture that are able to operate under Ice Age conditions. But will we do it? The task is enormous. Or will we fail ourselves on this front? We have no reason to allow us to fail. We have the materials and energy resources on hand to accomplish everything that is required for us to continue to live in an Ice Age World. But will we do it? The big question that never goes away, therefore, is; will we develop our inner resources as human beings sufficiently to get the job done, and to get it done in time? Or will we do nothing, ignore the challenge, and condemn our children and one-another to an agonizing death by starvation? That's the choice.

Towards meeting the inner challenge, I have created the epic series of novels, The Lodging for the Rose. And further, towards meeting the science challenge, I have produced numerous research books and several dozen exploration videos that the Illustrated Science series is modeled after. The work is the result of a quarter century of research, for which numerous elements of evidence in related fields came to light during the timeframe of my research.

It is my hope that the work that went into all of these projects will help in some degree - for humanity that we are all a part of - to write itself a ticket to have a future.

High-resolution color images, of the images in this book, can be obtained at www.iceagetheatre.ca

We have two models

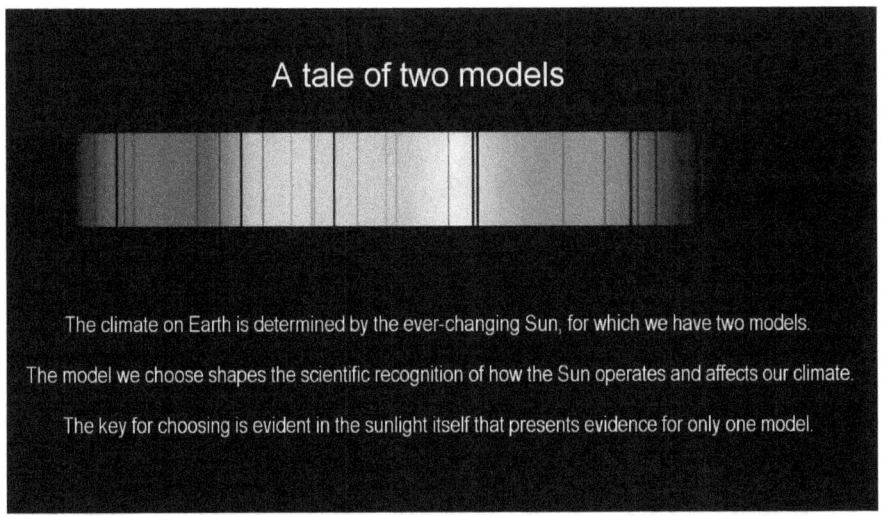

The climate on Earth is determined by the ever-changing Sun, for which we have two models.
The model we choose shapes the scientific recognition of how the Sun operates and affects our climate.
The key for choosing is evident in the sunlight itself that presents evidence for only one model.

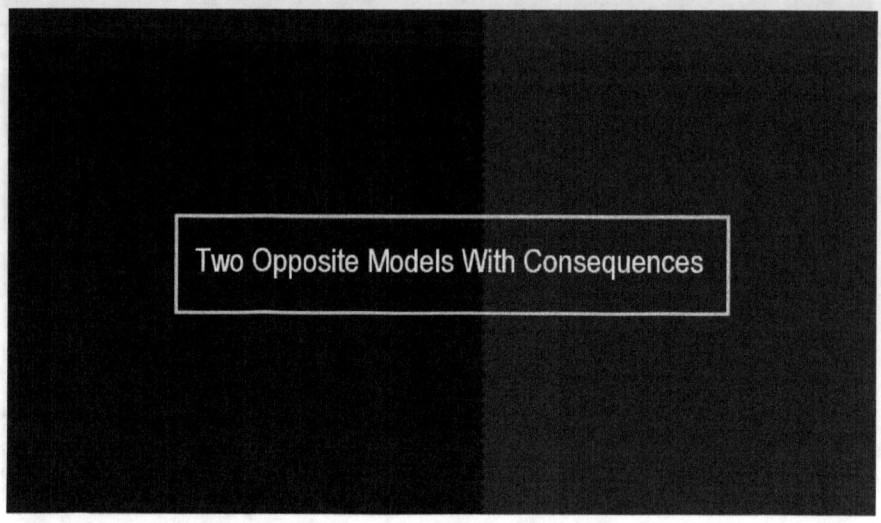

Two Opposite Models With Consequences

** Two Opposite Models with Consequences

Opposite models in the science of astrophysics

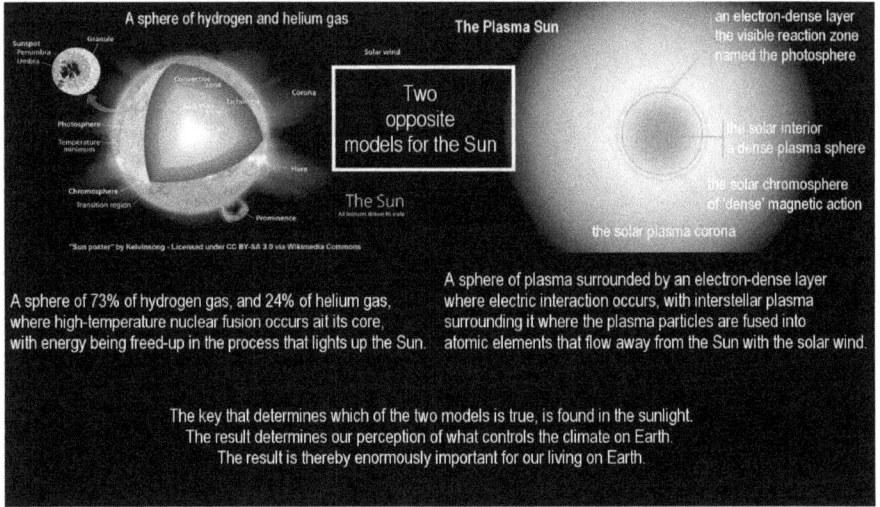

Two opposite models are being recognized in the science of astrophysics. One model presents the Sun as a sphere of 73% of hydrogen gas, and 24% of helium gas, where high-temperature nuclear fusion occurs at its core, according to the theory, by which hydrogen atoms are fused into helium atoms, with energy being freed-up in the process that lights up the Sun. That's the most-widely accepted theory.

The opposite model is that of the plasma-fusion Sun, that is a sphere of plasma surrounded by an electron-dense layer where electric interaction occurs with the interstellar plasma surrounding it, whereby plasma particles are fused into atomic elements that are electrically neutral and flow away from the Sun with the solar wind.

The real physical evidence tells us which model is represented by it.

Real physical evidence is often surprising

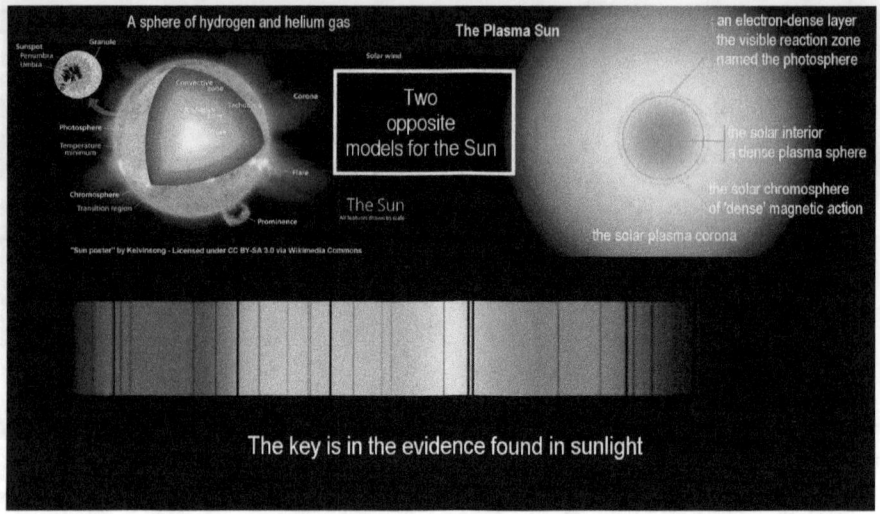

Which of the two models accords with the real physical evidence that we see in the sunlight, and which of the two models does not? Here the choice is not one of opinions, but is based on truth. Here the evidence begins to speak and stake its claim. The physical evidence tells us what is real. Of course, the real physical evidence is often surprising. It has the power to shatter illusions. Is this important? It is actually more important than one cares to acknowledge.

People react to what they perceive as reality

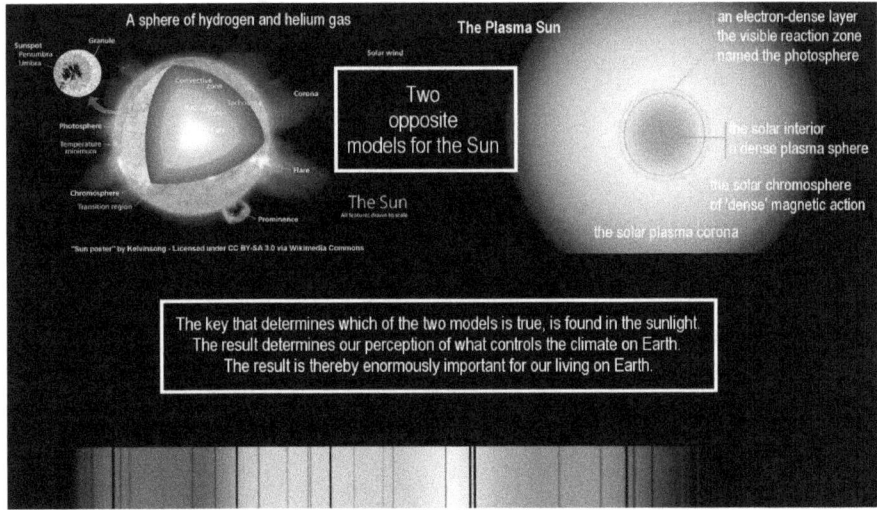

The manner in which people react to what they perceive as reality, which is so often based on opinion rather than evidence, determines their experiences. This includes the perception of what controls the climate on Earth. This perception has invariably, enormous consequences for our living on Earth as the result of our responding to our perceptions that are often just beliefs.
The two models that you see before you, represent total opposites. The model you choose, by the weight of the consequences, opens up issues for you that go far beyond merely academic considerations. It's the consequences that make the issues of extreme importance for your living on Earth. For this reason, I need to start here, with the consequences. This means I have to start with exploring what is inherent in each of the two models, in terms of the critical consequences. Doing this sets up the requisite stage for measuring the importance of what the evidence in the sunlight, as shown here, represents, which the video is designed to explore for its critical aspects.
So let's begin with a comparison of the two models in terms of their

consequences, as they are related to climate considerations.

*The Hydrogen Sun Model

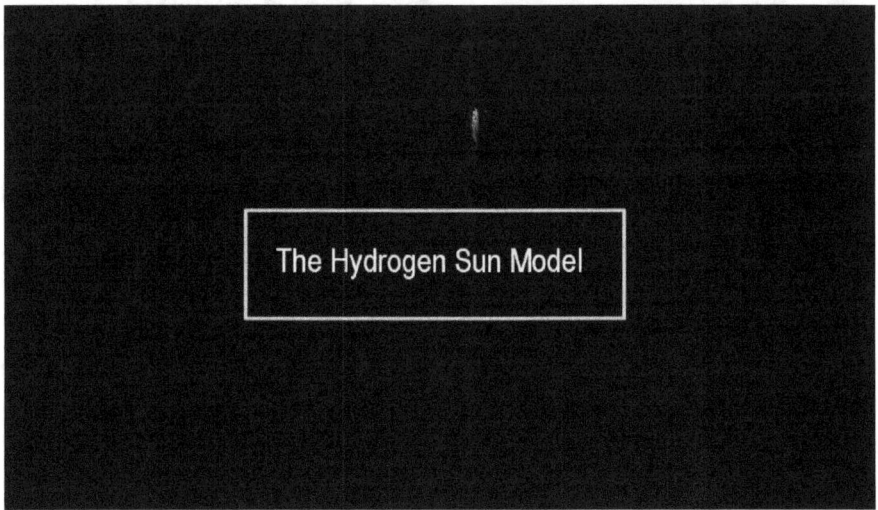

** The Hydrogen Sun Model

The internally powered hydrogen fusion Sun

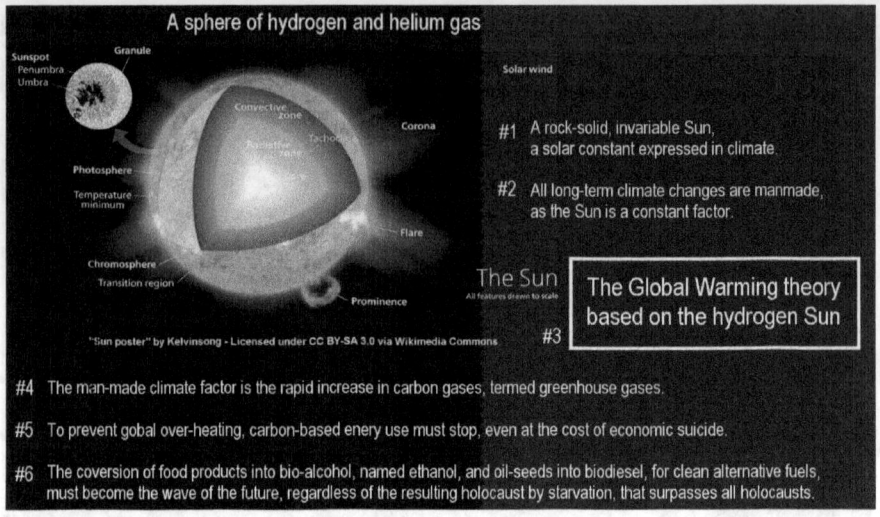

A sphere of hydrogen and helium gas

Sunspot
Penumbra
Umbra
Granule

Convective zone

Solar wind

Corona

#1 A rock-solid, invariable Sun,
a solar constant expressed in climate.

#2 All long-term climate changes are manmade,
as the Sun is a constant factor.

Photosphere

Temperature minimum

Flare

Chromosphere
Transition region

Prominence

The Sun
All features drawn to scale

#3

The Global Warming theory
based on the hydrogen Sun

"Sun poster" by Kelvinsong - Licensed under CC BY-SA 3.0 via Wikimedia Commons

#4 The man-made climate factor is the rapid increase in carbon gases, termed greenhouse gases.

#5 To prevent gobal over-heating, carbon-based enery use must stop, even at the cost of economic suicide.

#6 The coversion of food products into bio-alcohol, named ethanol, and oil-seeds into biodiesel, for clean alternative fuels,
must become the wave of the future, regardless of the resulting holocaust by starvation, that surpasses all holocausts.

The first model is the widely accepted model of the internally powered hydrogen fusion Sun.

#1 This model presents a Sun that is rock-solid in its operation, unvarying, a solar constant that one can count on. Under this model no changes in the energy intensity of the Sun is possible, or has occurred for millions of years.

#2 Under the model of the unvarying sun, all climate fluctuations on Earth are necessarily attributed to human action, for the perceived impossibility of any other cause affecting climate change.

#3 The human cause for climate change is attributed to mankind's emission of carbon gases, based on the observation that these gases have hugely increased in the global atmosphere since the start of the industrial revolution.

#4 Based on the assumed constant sun, and the measured manmade increase in atmospheric carbon gases, which are recognized to be greenhouse gases, the conclusion has been drawn that humanity has begun a course that will irreversibly heat up the Earth by trapping evermore radiated heat in the atmosphere, and thereby overheat the planet. Intolerable consequences are envisioned for such a future.

#5 In order to prevent the envisioned consequences, enormous demands are placed on humanity to radically shut down carbon based energy production all across the world, regardless of the resulting economic suicide that is already happening.

#6 The most-genocidal action, however, that is demanded on this front, is the presently near-worldwide-implemented biofuels mandate. The mandate has become a legal requirement for the nations to divert vast amounts of food products to be burned in cars in the form of ethanol and biodiesel, even at the cost of causing the worst holocaust in history of death by starvation

The mass-burning of food is murdering

Mass Murder with Biofuels
a YouTube video

El Tres de Mayo, by Francisco de Goya - Wikipedia

The enormous size of farm land that is taken out of food production, worldwide, for meeting the biofuels mandates, would normally nourish 400 million people. In a world that has a billion people living in chronic starvation, the mass-burning of food has become a holocaust by policy that is murdering, minimally, 100 million people per year by starvation in the greatest-ever genocide on Earth. It evidently takes a lot of corn, etc., to brew up 550 billion litres of alcohol hooch, which ethanol is, which is distilled up to 99% in solution to produce 92 billion litres of ethanol that is used to dilute gasoline with. The massive volumes of food that are sequestered from the nourishment of populations, to be burned for no appreciable benefit at all, are foods stolen from the most-needy.

The production of biofuels

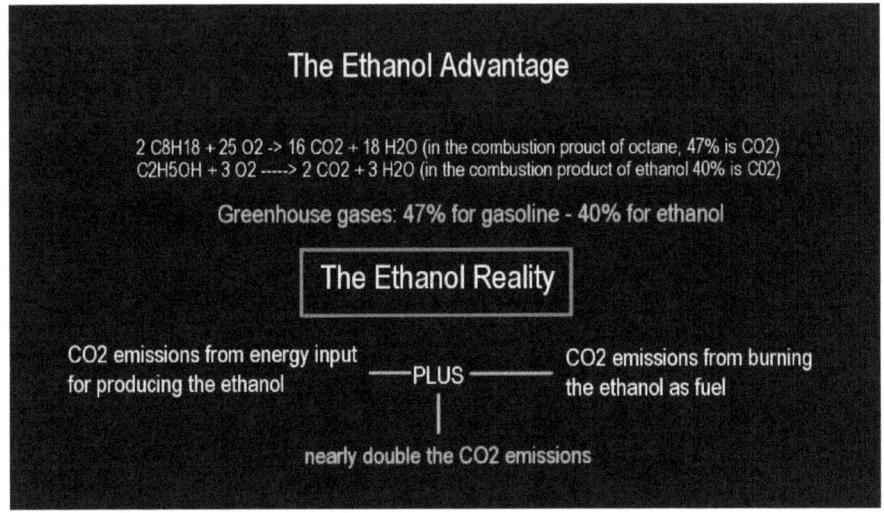

The production of biofuels, overall, requires nearly as much energy input from carbon sources, to produce the fuel, than the fuel gives back in energy. So it isn't an alternate energy resource. Why then would we bother? Nor is it the panacea of clean energy, that is free of greenhouse gases. From the burning of gasoline, for example, in the case of octane, 47% of the exhaust is CO_2. In the case of ethanol, the CO_2 exhaust is 40%. For the 7% difference, 100 million people are murdered each year with starvation.

$2\ C8H18 + 25\ O2\ ?\ 16\ CO2 + 18\ H2O$ (in the combustion product of octane,47% is CO2) $C2H5OH + 3\ O2\ -----\ 2\ CO2 + 3\ H2O$ (in the combustion product of ethanol 40% is C02)

In addition, the biofuels process, especially that for ethanol, with all the energy inputs factored in for producing the fuel, results in a fuel cycle that almost doubles the CO_2 emissions into the atmosphere for the applied energy use.

Thus, in the end the ethanol project violates the very objective that it was created for, which was supposed to reduce carbon emissions that are deemed greenhouse gases.

Biofuels depopulation genocide

british royal photography services

Victims of genocide in the Great imperial Famine of 1876–78 in India

40 famines in 3 centuries of British rule more than 58 million perished

From the Illustrated London News, (20 October 1877)

In short, the biofuels process is a scam that serves no other purpose than the depopulation genocide that has been spelled out in capital letters for many decades in numerous ways, as a chosen imperial policy. Fostering genocide has long been imperial policy, going back centuries in some cases, as in the case of India under British rule. The global warming issue is being used for this policy. This is easily done under the doctrine of the hydrogen Sun that is deemed to be rigidly invariable, as such a Sun would be if the concept was real.

The world has spent 40 years in climate dreaming

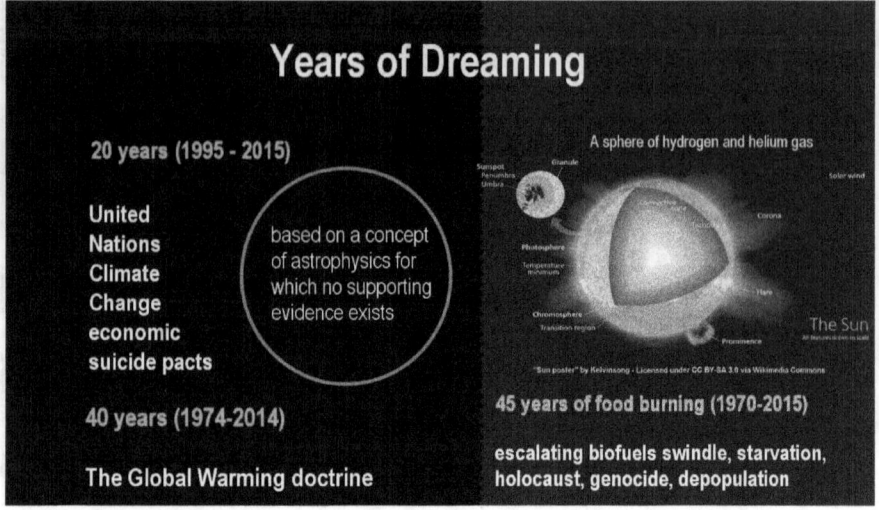

The world has spent 40 years in a state of dreaming, or climate dreaming as one might call it, with half of that time increasingly devoted to economic destruction and genocide. Ironically, the entire dreamscape and related tragedies, are rooted in a platform for which no actual real evidence exists, which opens up many avenues for dreaming. However, instead of dreaming, shouldn't one rather build the house of civilization on a model that is supported by real physical evidence?

*The Plasma Sun Model

** The Plasma Sun Model

The model for which real physical evidence exists

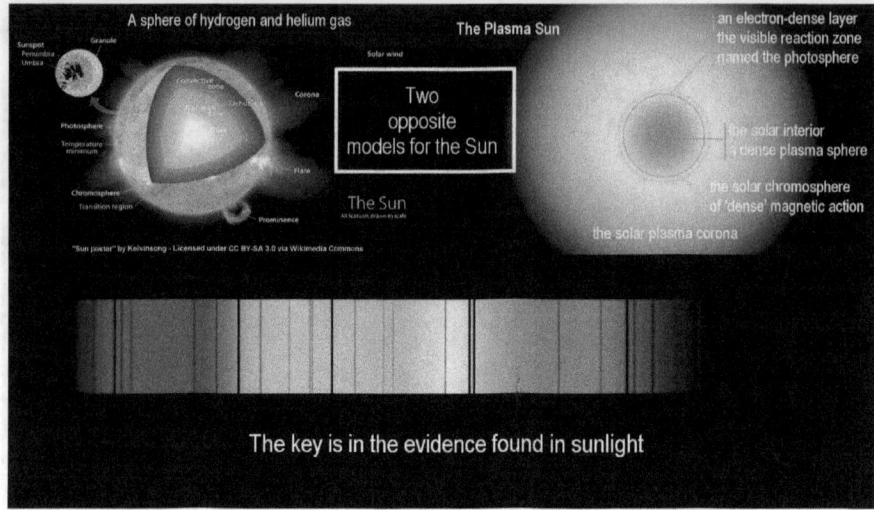

The model for which real physical evidence exists, wherever one looks, is ironically the opposite model to that which is generally accepted. The model that is supported by real physical evidence, is the model of the plasma Sun.

Here it gets interesting. There is a key found in the nature of the sunlight that is not possible on the platform of the hydrogen Sun, which thereby disproves it, but which is inherently natural on the platform of the plasma Sun.

The plasma Sun has a totally different basis

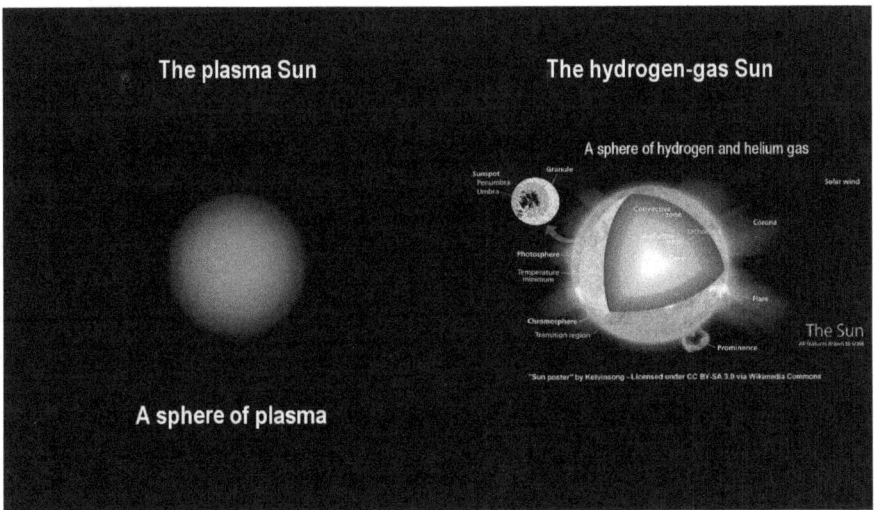

The plasma Sun has a totally different basis than that theorized for the hydrogen-gas Sun model. Plasma is not a gas. It is not atomic in nature. It is not visible to the eye. It consists of Protons and Electrons that are the electrically charged building blocks that all the atoms in the universe are assembled from. Plasma is too small to be visible. The larger type of the plasma particles is 100,000 times smaller than the smallest atom, and the electron is 1000 times smaller than that.

Electric charge that is 39 orders of magnitude stronger

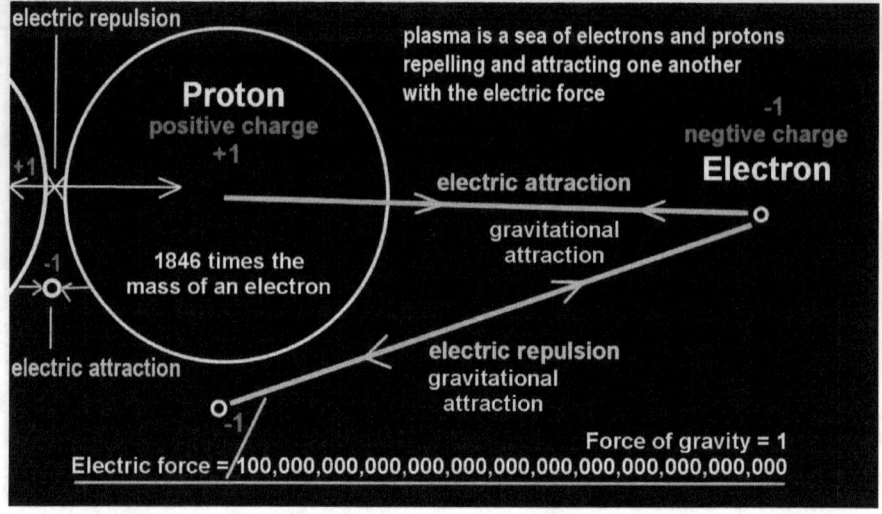

Though being small, the plasma particles carry an immense electric charge that is 39 orders of magnitude stronger than the force of gravity. The Proton has a positive charge, and the Electron a negative charge. By the electric force, particles of like charge repel each other, and of unlike charge attract one another. By the protons repelling one-another, free plasma in space tends to be highly diffused. This means that a plasma Sun has a 1000 times lower mass-density than a hydrogen Sun of the same size would have if such a phenomenon was possible.

Unlike electric charges attract each other

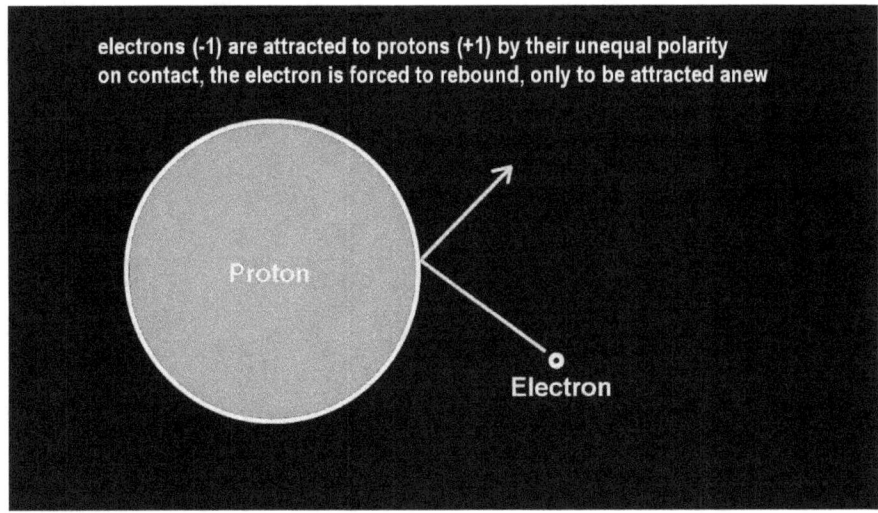

Furthermore, unlike electric charges attract each other. By this electric force, electrons, with 1800-times less mass, are always attracted to protons, though they never latch on. Before the two can latch together, a feature of the electric force causes the electron to be repelled at a close distance, whereby it rebounds into what resembles an endless dance.

Electrons migrate away from the center of gravity

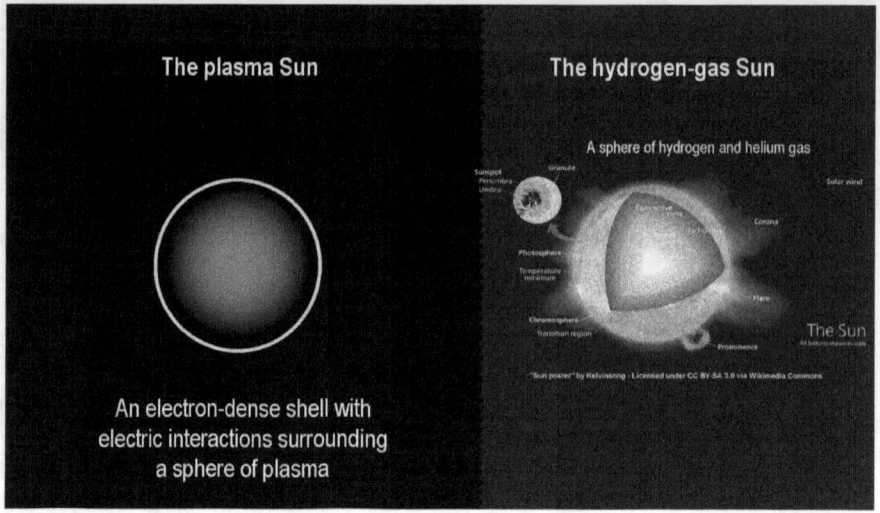

In the soup of mutually repelling protons and dancing electrons, the dancing electrons that are 1800 times lighter than the protons, tend to migrate away from the center of gravity where they form an electron-rich layer that we behold as the surface of the Sun. We recognize it as a surface by the products of the reactions that occur there.

A plasma Sun is powered from the outside

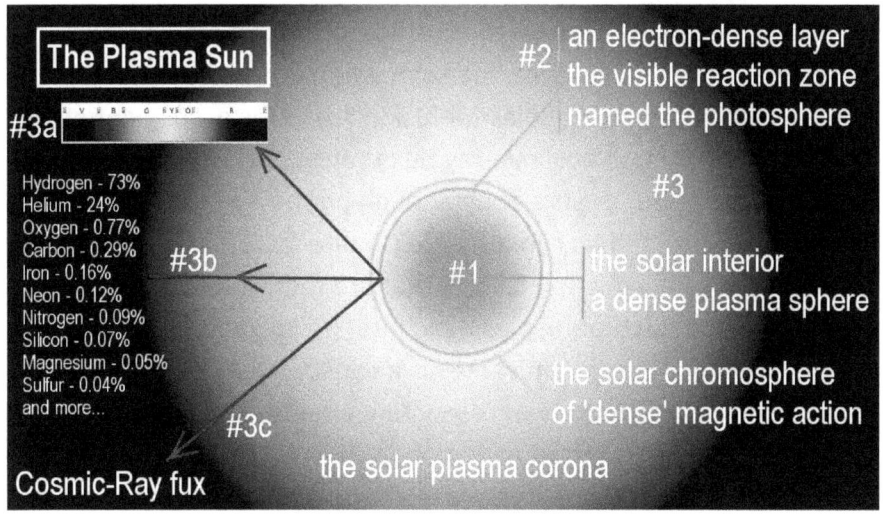

A plasma Sun has many features.

#1 As I said before, at the core is a sphere of cosmic plasma that is held together by gravitational force.

#2 As I also said, gravitational pressure is acting on the plasma, which forms an electron-dense layer that becomes its 'surface,' where the Sun becomes electrically active.

#3 This means that a plasma Sun is not internally powered, but is being powered from the outside by interstellar plasma being focused onto it from afar. The interaction, that is electrically accelerated, causes plasma-fusion reactions to occur at the solar surface.

In the resulting high-energy fusion process at the surface of the Sun, three major types of emissions become possible that are inherent in the physics of a surface-active Sun.

#3a The most visible effect is the sunlight. A wide spectrum of sunlight and solar energy is produced in the electric fusion process that gives us all the colors in a continuous band. This feature is not possible on any other basis.

#3b The feature, that makes the sunlight possible, is the plasma-fusion process itself, in which all the known natural atomic elements are synthesized on the surface of the Sun, which subsequently flow away with the solar wind.

#3c Since the high-energy plasma-fusion reactions occur right at the surface of the Sun, the cosmic-ray flux that high-energy plasma reactions generate, exists strong enough on the Sun to significantly impact the Earth and its climate. The climate factors on Earth are caused by solar cosmic-ray factors that are themselves caused by resonating and fluctuating plasma in interstellar space. There is no such thing in the plasma universe, as a constant Sun. Electric plasma is subject to a wide variety of resonating effects, which affect the Earth secondarily, in the form of climate fluctuations. There is no such thing as a constant climate on Earth, or ever has been in all of history. The ice core evidence makes this fact rather plain.

Huge climate fluctuations

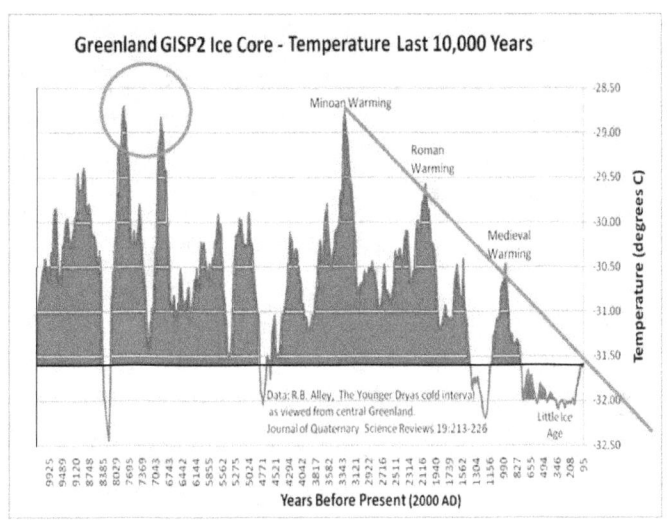

Look at the huge climate fluctuations at about 8300 years ago, which dwarves the little Ice Age climate-change. This wasn't caused by manmade greenhouse gases. The world population was minuscule then.

Also look at the enormous down-ramping of the global climate from the Minoan Warming, to the Roman Warming, and the Medieval Warming, right to the present world. This enormous down-ramping, with huge fluctuations along the way, are features that one would expect to see as the result of resonating plasma in interstellar space that powers our Sun.

What you see here wasn't caused by human effects, much less by human contributions to greenhouse gases, and even less so by CO2 that isn't a factor at all, except in the imagination.

*The Big Climate Factors

** The Big Climate Factors

Cosmic dynamics that affect our Sun

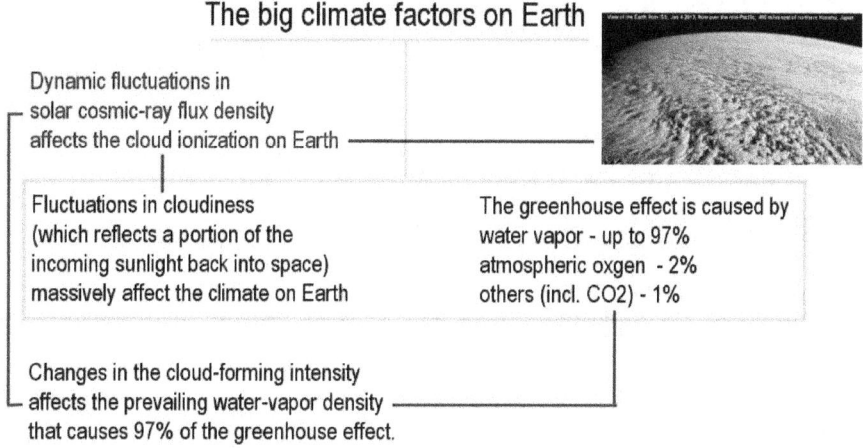

The big climate factors on Earth

Dynamic fluctuations in
solar cosmic-ray flux density
affects the cloud ionization on Earth

Fluctuations in cloudiness
(which reflects a portion of the
incoming sunlight back into space)
massively affect the climate on Earth

The greenhouse effect is caused by
water vapor - up to 97%
atmospheric oxgen - 2%
others (incl. CO2) - 1%

Changes in the cloud-forming intensity
affects the prevailing water-vapor density
that causes 97% of the greenhouse effect.

When up to 97% of the greenhouse effect of the Earth's atmosphere is caused by water vapor, which is hugely affected by effects on the Sun, eyebrows should be raised against the claims of the CO2 climate forcing. Furthermore, when as little as a millionth part of the greenhouse effect is demonstrably caused by CO2, of which the manmade contribution is roughly 10%, then the atmospheric CO2 concentration, no matter how little or how large it may be, is insignificant for comparisons with the ever-changing cosmic dynamics that affect our Sun, which hugely affect the Earth and its climate with solar cosmic-ray-flux fluctuations.

With plasma streams being electric in nature, electric resonance effects occur at all levels, with the result that the plasma density around the Sun is constantly changing. In other words, the Sun is not a constant factor for the climate on Earth. Enormously large effects of the Sun affect the climate on Earth in huge ways, by affecting cloudiness and water-vapor density in the atmosphere, which are two factors that massively affect our climate.

The Solar Cosmic-ray factor is so powerful

View of the Earth from ISS, Jan.4 2013, from over the mid-Pacific, 460 miles east of northern Honshu, Japan.

In comparison with the ever-changing cosmic dynamics that affect our climate, carbon gases are insignificant. The Solar Cosmic-ray factor is so powerful in affecting cloudiness that reflects a portion of the incoming solar energy back into space, which is thereby lost to us, that in comparison, the human-activity factors amount to nothing. They truly affect nothing at all, including CO_2. Only the ghost of CO_2 affects us, as a mythological factor, as it conjures up dream visions where nothing is actually real. The most critical need in modern times is, for society to open its eyes to what is demonstrably real as founded in verifiable physical evidence and understandable universal principles.

CO2 a millionth part of the greenhouse effect

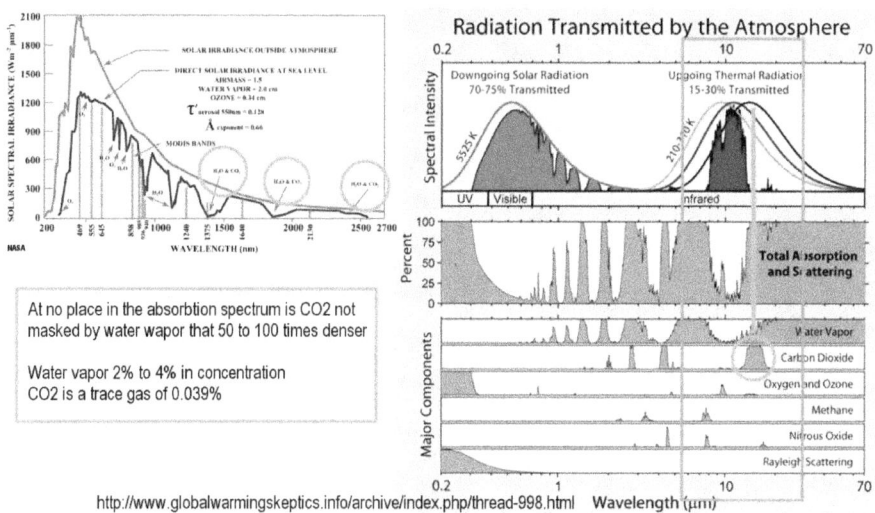

At no place in the absorbtion spectrum is CO2 not masked by water wapor that 50 to 100 times denser

Water vapor 2% to 4% in concentration
CO2 is a trace gas of 0.039%

http://www.globalwarmingskeptics.info/archive/index.php/thread-998.html

Since the changing cloudiness also affects the atmospheric water vapor density that causes up to 97% of the greenhouse effect according to available evidence, the CO2-density factor is too small a factor to affect anything in this gigantic arena where cosmic factors run the show.

As I said before, while the CO2 has the characteristic to qualify it as a greenhouse gas, its minuscule density in the air, and the fact that its effect is masked almost completely by the vastly more-powerful water-vapor effect, enables the CO2 to contribute only roughly a millionth part of the overall greenhouse effect of the atmosphere of the Earth.

Since the accumulated human contribution to the global atmospheric CO2 since the start of the industrial revolution, adds up to a mere 10% of it, the result has no effect on the climate that is of practical significance. In the real world, CO2 is not a climate factor. The solar cosmic-ray flux is the climate factor that overrides all other factors. However, the 10% increase in atmospheric CO2 that humanity has caused since the start of the industrial revolution, has a significant effect on increasing plant growth. But this is a different subject, and a big subject all by itself.

Atmospheric CO2 is a good thing for us all

As I had pointed out in several other videos, the increasing atmospheric CO2 is a good thing for us all. It makes the world greener, and agriculture more productive. We need more atmospheric CO2 for all the green plants to thrive, not less. The world is presently so starved of CO2, that when greenhouse operators double the CO2 concentration in their facilities, a 50% increase in plant growth results.

The current CO2 deficiency is the really big critical issue for all life on Earth. It's bigger than the hyped-up global warming myth.

The CO2 deficiency promises to become super-critical

The CO2 deficiency promises to become super-critical when the next Ice Age begins, which will likely happen in 30 years according to a wide range of credible evidence.

The Plasma-Sun model tells us

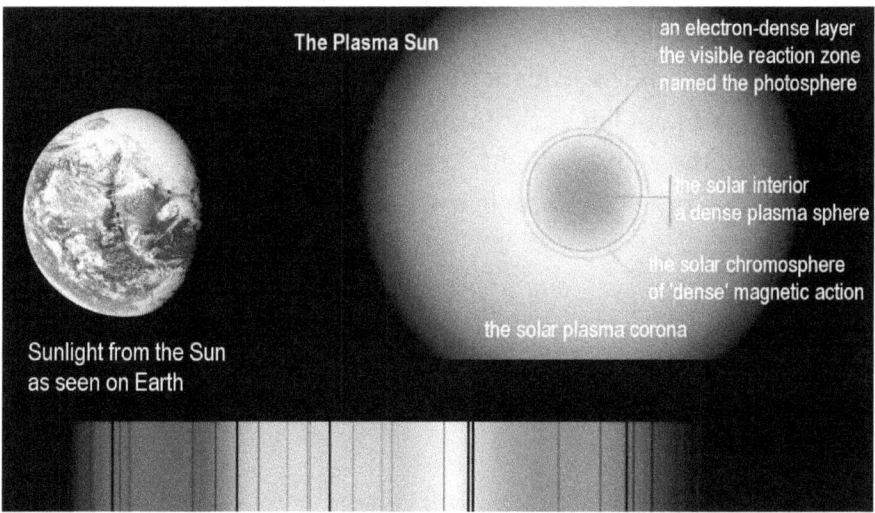

And all of this, including the Ice Age dynamics, is deeply related to what we see evident in the sunlight, because what we see there, indicates which type of solar model is supported by real evidence, which in turn affects the consequences that we experience.

The evidence that we see in the sunlight here is only possible on the platform of the plasma Sun, where cosmic-ray factors cause all climate changes on Earth. The Plasma-Sun model tells us that all the draconian acts to prevent global warming can be safely scrapped, and that committing economic suicide to prevent global warming is totally unnecessary. Likewise, can the biofuels holocaust in the name of preventing global warming, be scrapped then, that is presently murdering 100 million people a year with starvation, which is simply not needed then - which is a crime against humanity from the beginning. It also means that the global CO_2 levels can be safely increased 10-fold to historic levels to invigorate the biosphere.

With the plasma-sun model becoming recognized, humanity gives itself a foundation that allows it to live again and develop, and to be free of all depopulation ideologies. But is the evidence that we see in the sunlight really that big, so that it enables such a wide liberation on all fronts?

The answer is yes. The evidence is that big. The key is found in what the numerous faint lines represent that are spangled all across the spectrum of the white sunlight. These faint lines have a powerful cause that no other platform in the universe, than the plasma Sun, can create.

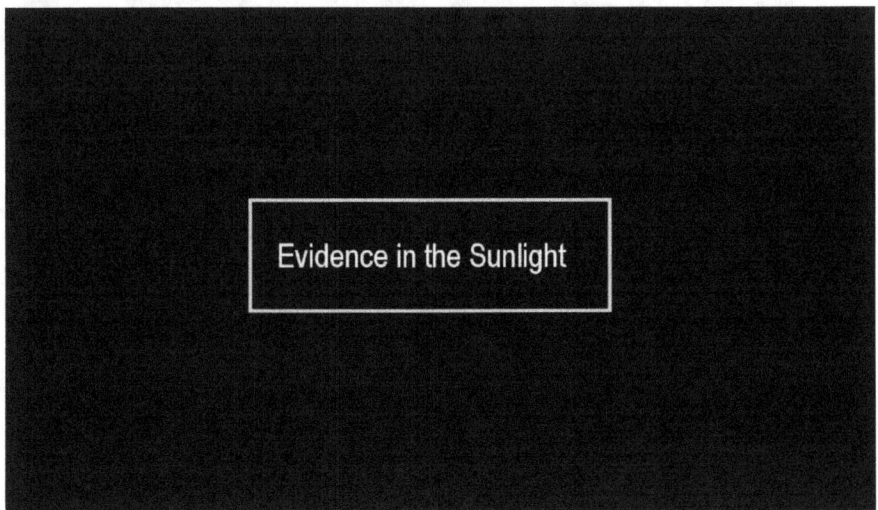

** Evidence in the Sunlight

Absorption lines

These faint lines are generally referred to as, absorption lines.

The Light Absorption Principle

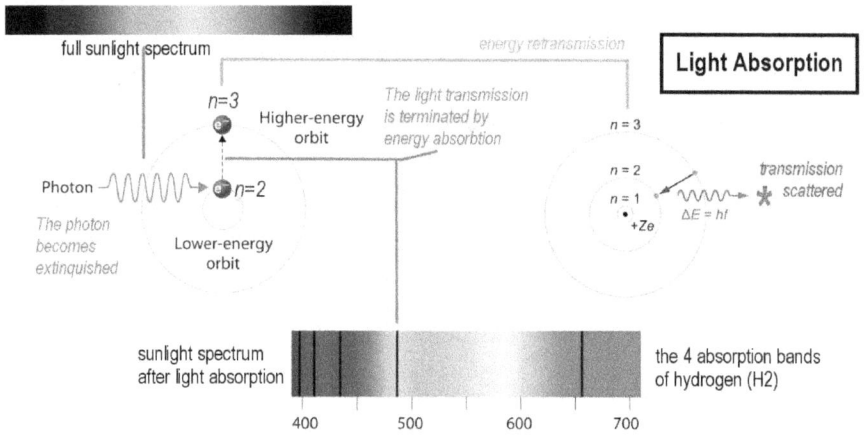

The Light Absorption Principle

Light absorption occurs when photons of a specific energy level encounter atoms in their path that absorb light of specific wavelengths.

Photons are packets of energy that are the carriers of what we call light. The photons come in different sizes with different energy contents that are specific to the different colors. When a matched encounter of a photon and an atom happens, the energy that is carried by the photon is captured and absorbed by an electron of the atom that the photon encounters.

In this process, the photon ceases to exist. The absorbed energy forces the electron into a higher orbit that corresponds with the electron's higher energy-state. Since these energized electrons don't fit into the makeup of the original atomic structure, the electron reverts back to its original domain in the form of a type of electromagnetic wave within the atom. When this process happens, the previously absorbed energy forms a new photon.

The new photon that is born in this process, however, has no specific directionality. The re-transmission of absorbed light is scattered into all directions. This means that to the observer of the sunlight, the absorbed light-stream is lost.

Since the entire process of light absorption and retransmission is specific to the makeup of an atomic element, the absorbed color bands are specific to an atom's internal structure. Some atomic elements are able to absorb several specific colors, as in the case of the hydrogen atom shown here.

The numerous absorption lines

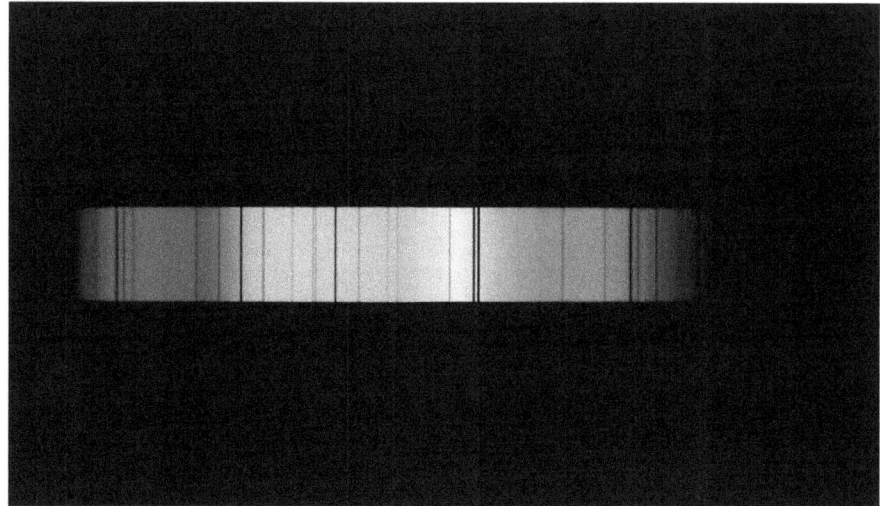

The numerous absorption lines that you see here in sunlight as it is observed on the surface of the Earth, tell us that the sunlight has encountered numerous different types of atomic elements on its path to us, which is evident by the numerous absorption lines spread all across the white-light spectrum. Evidently, the absorption lines were caused AFTER the sunlight was created. They were not in the original sunlight. They tell us of large volumes of specific atomic elements that the light from the Sun has encountered on the way to us.

The discovery process for all this began in the 1800s.

The Fraunhofer Lines 1814-1815

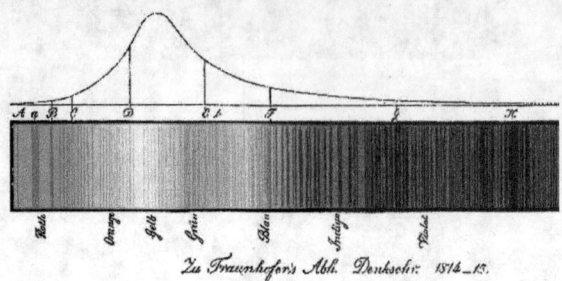

It was recognized in 1802, by the English chemist William Hyde Wollaston, that a number of dark features appear in the sunlight spectrum. A dozen years later, in 1814, the German physicist Joseph von Fraunhofer (1787–1826) independently discovered these features in the form of slightly darker lines appearing in various places in the sunlight spectrum. He didn't know what they mean, but he studied them, and mapped them. He did an exhaustive study of them, in which he carefully measured their wavelength by their location in the spectrum. In this manner he mapped 570 of such lines. In honor of this work, these lines were named the Fraunhofer Lines.

Fraunhofer's lines coincide with emission lines

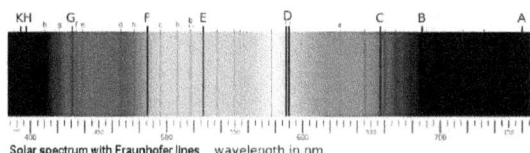

Solar spectrum with Fraunhofer lines   wavelength in nm

Fraunhofer designated the principal features with the letters A through K, and the weaker features with other letters.
About 45 years later it was noted by the researchers Kirchhoff and Bunsen that a number of Fraunhofer's lines coincide with the discovered characteristic emission lines identified by the spectra of heated elements, such as in the case of hydrogen. It was deduced by them that the dark features in the solar spectrum are caused by light-absorption of chemical elements existing in the Solar atmosphere.

The Balmer Series

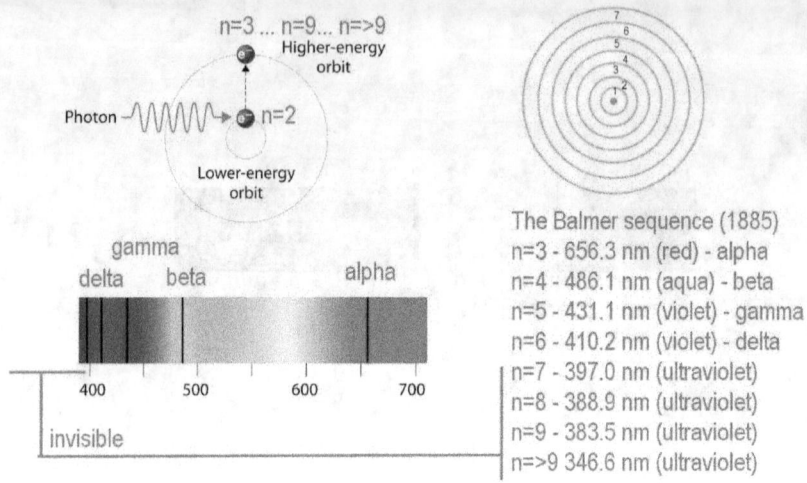

The Balmer sequence (1885)
n=3 - 656.3 nm (red) - alpha
n=4 - 486.1 nm (aqua) - beta
n=5 - 431.1 nm (violet) - gamma
n=6 - 410.2 nm (violet) - delta
n=7 - 397.0 nm (ultraviolet)
n=8 - 388.9 nm (ultraviolet)
n=9 - 383.5 nm (ultraviolet)
n=>9 346.6 nm (ultraviolet)

The Balmer Series
Swiss mathematician and mathematical physicist, Johann Jakob
Balmer, discovered in 1885 the existence of a lawful geometric
progression of the spectral lines of the hydrogen atom. Balmer
created a mathematical series based on successive energy quantum
states of an electron in an atom, that corresponds to successively
larger transitions away from the quantum base of 2, in his series. He
extended the progression all the way to 9 and beyond. The
computed values, of course, are also true in reverse for the
emission spectral lines of light emitted when the electron converts
its excess energy into a new photon as it resumes its normal
position in the atom.
Why is this significant?

Light energy absorption of chemical elements in the space

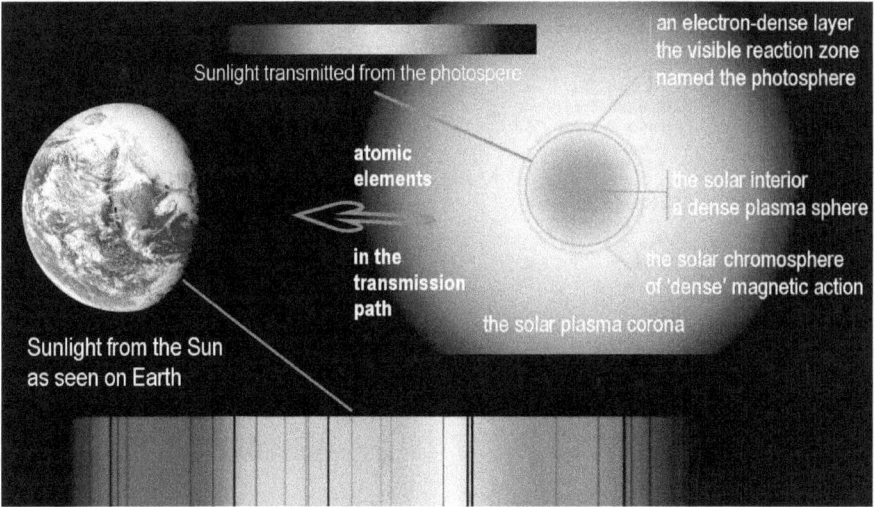

It is significant as proof that the spectral lines that Fraunhofer had discovered and mapped are the lawful results of light energy absorption of chemical elements in the space between the photosphere where the sunlight is created, and the observer on the Earth.

Since the different structures of different atoms result in different, but specific absorption lines which have been recognized as their signature, it becomes possible to determine with great certainty what specific atomic elements are present in high concentrations between us and the source of the sunlight. The result is significant, because what we see it is not what one would expect.

The main signature lines

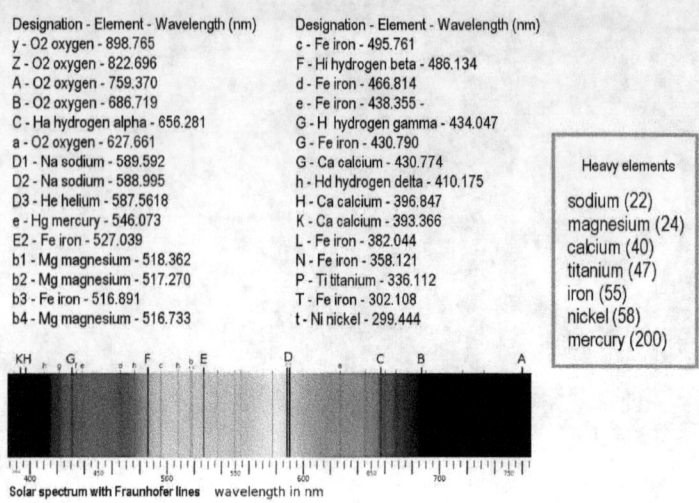

Designation - Element - Wavelength (nm)
y - O2 oxygen - 898.765
Z - O2 oxygen - 822.696
A - O2 oxygen - 759.370
B - O2 oxygen - 686.719
C - Ha hydrogen alpha - 656.281
a - O2 oxygen - 627.661
D1 - Na sodium - 589.592
D2 - Na sodium - 588.995
D3 - He helium - 587.5618
e - Hg mercury - 546.073
E2 - Fe iron - 527.039
b1 - Mg magnesium - 518.362
b2 - Mg magnesium - 517.270
b3 - Fe iron - 516.891
b4 - Mg magnesium - 516.733

Designation - Element - Wavelength (nm)
c - Fe iron - 495.761
F - Hi hydrogen beta - 486.134
d - Fe iron - 466.814
e - Fe iron - 438.355 -
G - H hydrogen gamma - 434.047
G - Fe iron - 430.790
G - Ca calcium - 430.774
h - Hd hydrogen delta - 410.175
H - Ca calcium - 396.847
K - Ca calcium - 393.366
L - Fe iron - 382.044
N - Fe iron - 358.121
P - Ti titanium - 336.112
T - Fe iron - 302.108
t - Ni nickel - 299.444

Heavy elements

sodium (22)
magnesium (24)
calcium (40)
titanium (47)
iron (55)
nickel (58)
mercury (200)

Solar spectrum with Fraunhofer lines   wavelength in nm

The main signature lines reveal the presence of a number of
extremely heavy elements existing in the Sun's corona, such as
sodium, magnesium, calcium, nickel, iron, titanium, even mercury,
with mercury being nearly as heavy as lead. Such extremely heavy
elements shouldn't exist in the solar atmosphere in such great
quantities that highly visible absorption bands result from their
presence. If the Sun was a hydrogen sun, gravitational attraction
should have drawn these heavy elements deep within the Sun
millions of years ago. But this is not what we see.
The answer is simple. The heavy elements that have been detected
to exist in great abundance in the Sun's corona, exist there, because
they are constantly being created by the Sun. The have a dynamic,
not a static, presence. They are synthesized in plasma fusion in the
photosphere, and are distributed away from there with the solar
winds. This is the only possibility for them to be existing there.
The presence of these heavy elements in the corona is not a static
phenomenon, like the atmosphere of the Earth. Their presence is
dynamic. They are in motion. They are out-flowing from the Sun.

They are synthesized on the surface of the Sun, in a process of electric plasma fusion. They flow through the corona with the solar wind. This is what the Fraunhofer lines are telling us. This is also the only possibility for the Fraunhofer lines to exist. They reflect the existence of vast volumes of atomic elements in the solar corona and in the space around it, as they flow away while they are constantly replenished with the plasma-fusion process that lights up the Sun.

Fraunhofer lines are caused right at the Sun

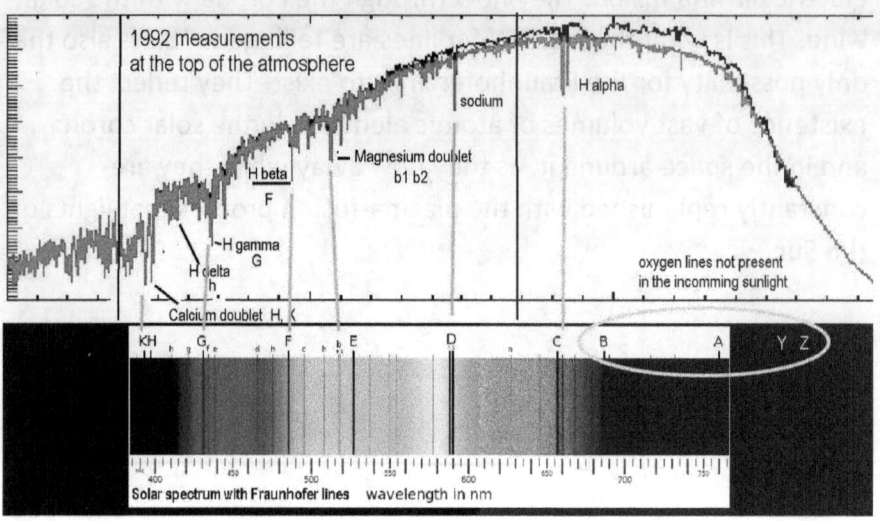

Solar spectrum with Fraunhofer lines   wavelength in nm

That the Fraunhofer lines are caused by absorption of sunlight right at the Sun, and is not the result of light absorption in the Earth's atmosphere, becomes evident when we look at the sunlight at the top of our atmosphere. We see the major light absorption spectra clearly evident there, even without contrast enhancement.

The only absorption spectra that are missing in the sunlight above the Earth's atmosphere, are the spectra that are specific to oxygen. A whopping 23% of the Earth's atmosphere is oxygen. The oxygen related absorption spectra are clearly evident to have been added on Earth by sunlight passing through our oxygen-dense atmosphere. The oxygen lines are named for this reason, the 'telluric lines.' This means that the remaining absorption lines were caused right at the Sun. This also means that our Sun is demonstrated to be a plasma Sun, which means that no other type of Sun really exists.

Fraunhofer lines stand as evidence for the plasma Sun

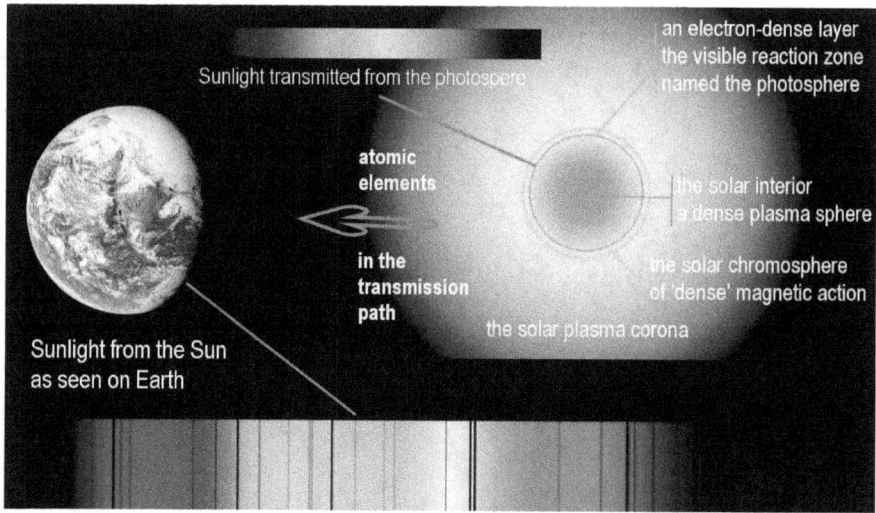

The Fraunhofer lines also stand as evidence for the plasma Sun in still another way. They tell us in a clearly measurable fashion that extremely heavy atomic elements exist in the solar corona, such as mercury that is almost 4 times heavier than iron. With the mercury having a dynamic existence in the corona, like all the other elements that cause absorption lines, the question needs to be asked, what happens to them, especially the heavy mercury? Do we have evidence of its falling out from the solar wind? The answer is yes.

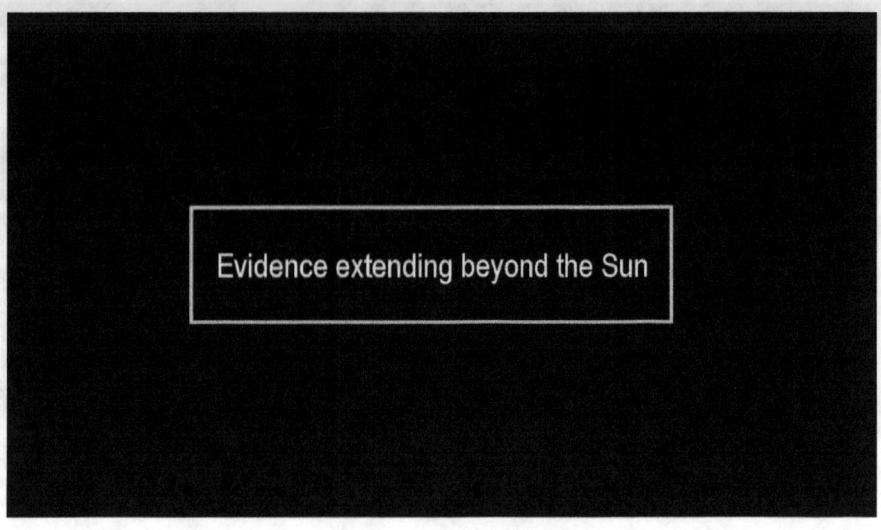

** Evidence extending beyond the Sun

The Planet Mercury Paradox

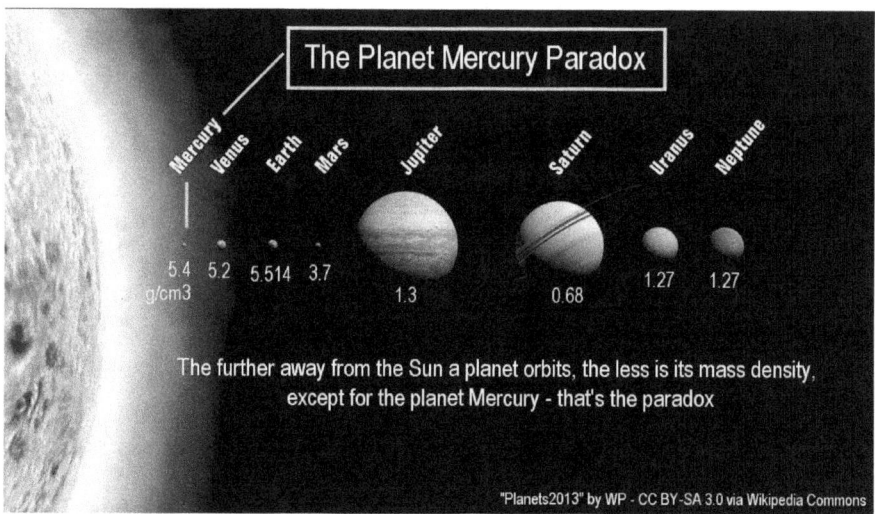

The Planet Mercury Paradox

The further away from the Sun a planet orbits, the less is its mass density, except for the planet Mercury - that's the paradox

"Planets2013" by WP - CC BY-SA 3.0 via Wikipedia Commons

The Planet Mercury Paradox

The further away from the Sun a planet orbits, the less is its mass density, except for the planet Mercury - that's the paradox.

The evidence does exist. It actually solves one of the puzzles in astronomy that one might call the Mercury paradox. In relationship with the Earth's mass density, Mercury's mass density is the second highest in the Solar System and only slightly less in total than that of the Earth.

Gravitational compression factored out

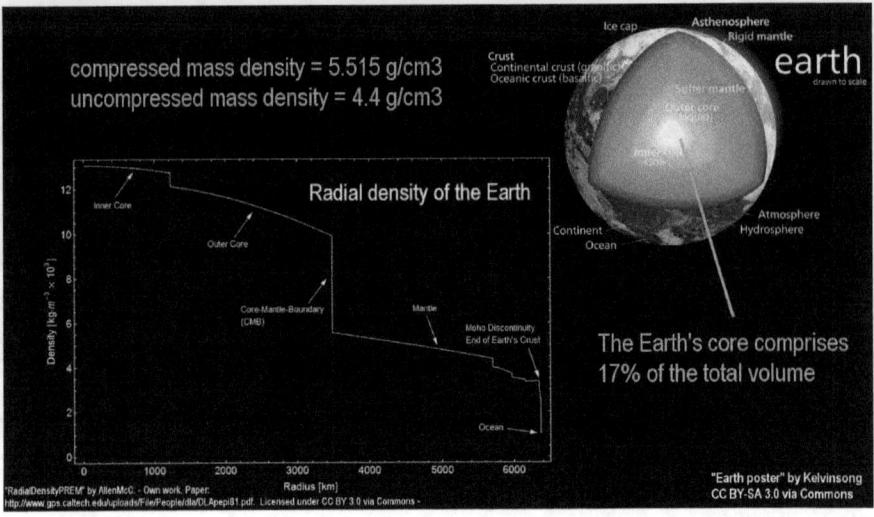

But Mercury is a tiny planet that has far less gravitational compression weighing on its material than on the Earth. It has been calculated that if the effect of gravitational compression were to be factored out of the mass-density comparison, the materials of which Mercury is made, would be significantly denser than that of the Earth.

Theories to bridge this self-evident paradox

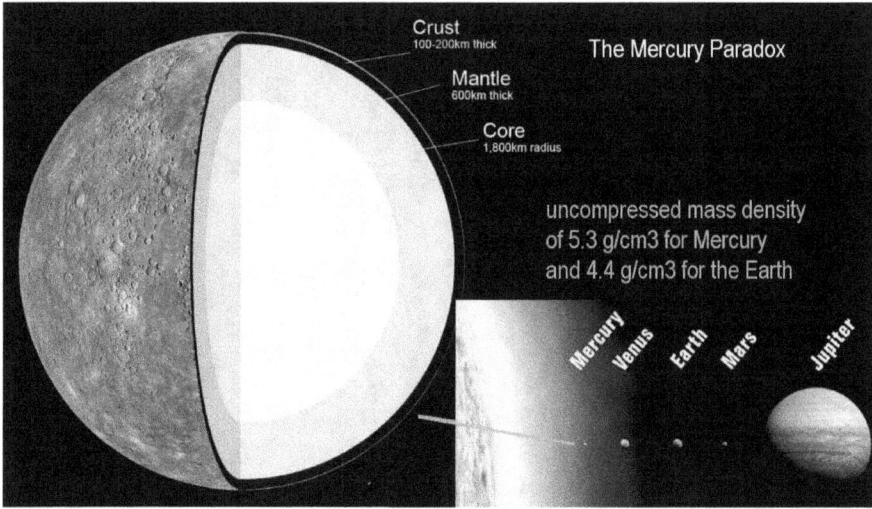

Mercury would have an uncompressed mass density of 5.3 g/cm3 versus the Earth's 4.4 g/cm3. Mercury's mass density would then be 20% greater, than that of the Earth. This is hard to rationalize for such a small planet.

Exotic theories have been invented to bridge this self-evident paradox. One of the solutions that has been proposed is to assume that Mercury's metallic core occupies about 42% of the planet's volume, as a sphere of iron 3,600 Km in diameter. This compares with 17% for the Earth.

A much more likely solution for the Mercury paradox would be, to simply acknowledge that with Mercury being the closest planet to the Sun, it would have received the heaviest materials falling out from the solar wind during the time when the solar system was formed. This means that a part of its core would contain large amounts of the more-heavy metals, such as mercury, platinum, and Gold, which the Fraunhofer lines have demonstrated in principle are flowing off the Sun.

Solar system as evidence for the plasma Sun

"Planets2013" by WP - CC BY-SA 3.0 via Wikipedia Commons

In this manner the entire solar system itself, stands as evidence for the plasma Sun, with Mercury having the highest basic mass-density, by being the closest planet to the Sun.

This principle of decreasing mass-density in the planets the further a planet is located away from the Sun, is evident throughout the solar system.

The diminishing mass-density to progressively lower values, reflects the fact that the lighter materials are carried further in the solar wind, with less fallout along the way. Only the frozen gas planets, Neptune and Uranus don't match this progression. Being the coldest planets in the range of minus 200 degrees Celsius, they weigh in with a larger mass density than Saturn, but less than Jupiter. Nevertheless, the unique overall mass-density distribution across the solar system delivers clear proof for the plasma Sun model as being totally real, which of course, disproves the hydrogen Sun model.

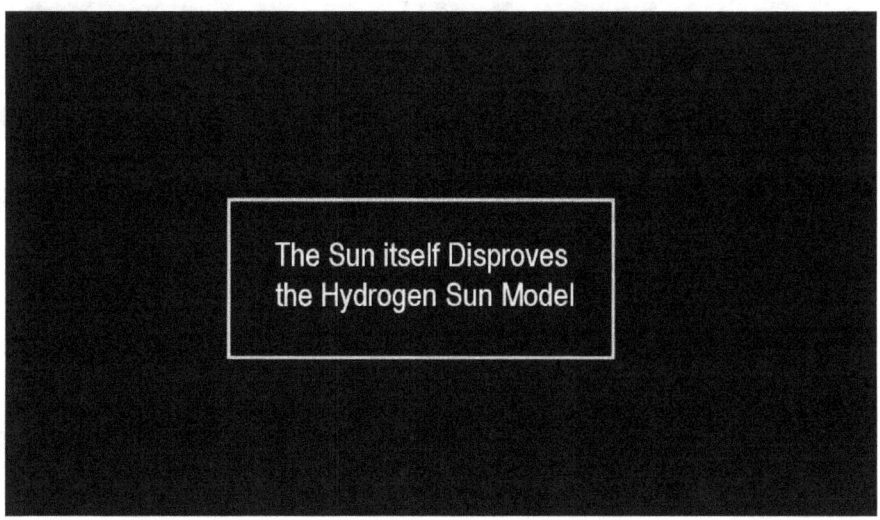

** The Sun itself Disproves the Hydrogen Sun Model

Sunlight delivers rigorous proof for the plasma Sun

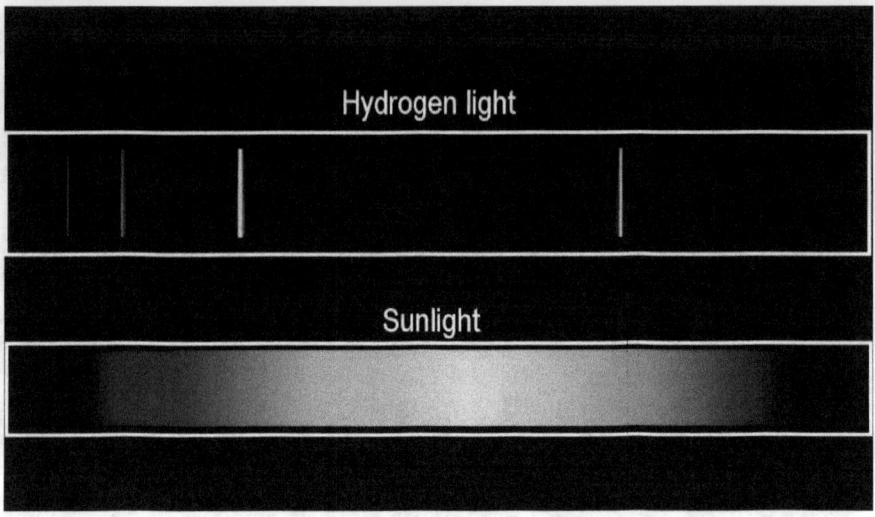

A similar paradox that is evident in the sunlight, likewise delivers rigorous proof for the plasma Sun model, which again, thereby solves the paradox. The paradox is the white sunlight itself as a continuous band of color that spans the entire visible spectrum and beyond. The hydrogen Sun model is physically incapable of producing the full spectrum of colors that we see. Just as the hydrogen atom can absorb light in only four narrow bands of the visible spectrum by the possible quantum mechanics of its electron's orbital position, so is the hydrogen atom only capable of emitting light in the same narrow bands by the actions of the same quantum mechanics. This means that a hydrogen Sun is simply incapable of emitting the full spectrum of the white sunlight. The hydrogen Sun model is thereby disproved.

The heavy elements of Oxygen, Carbon, Iron, Neon, Nitrogen, Silicon, Magnesium, and Sulfur, that have been detected in the Sun's atmosphere, would not exist there if the Sun was a hydrogen Sun. These heavy elements would have all been drawn into the core, by gravitational forces, leaving only hydrogen atoms to produce the sunlight. Their presence on the Sun, prove the sun to be a plasma Sun.

Hydrogen Sun is a 1000-fold too light

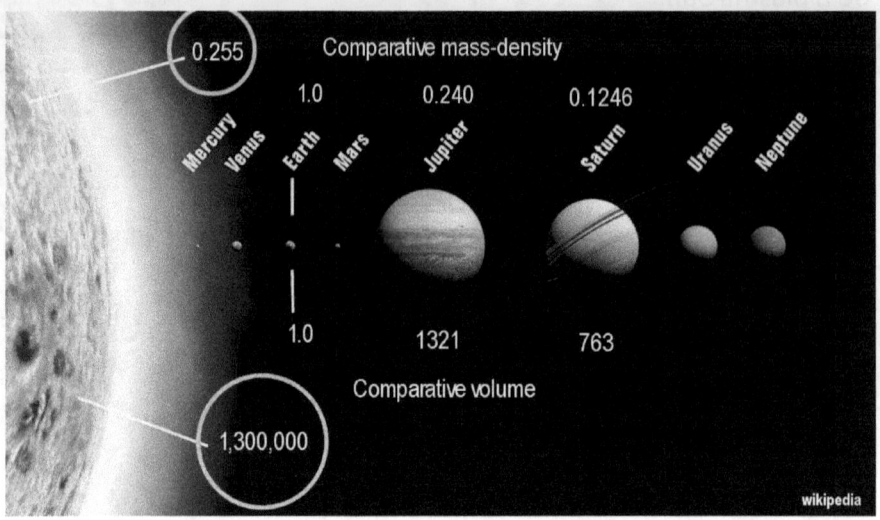

Another hydrogen Sun paradox is one that is related to the Mercury mass-density paradox. For its size, as a sphere of hydrogen and helium gas, the Sun is a 1000-fold too light. If one compares the volume and the mass density of Saturn and Jupiter, it becomes apparent that Jupiter, with double the volume, also has roughly double the mass-density as one would expect to result from the greater gravitation compression of the gases of Jupiter. If one adds the Sun to the comparison, which has thousand times greater volume, the corresponding mass-density increase by the greater gravitational compression, simply doesn't exist. The recognized mass-density for the Sun is roughly the same as Jupiter. This large discrepancy is a paradox. It shouldn't exist, but it does exist. The paradox is solved when the Sun is understood as a sphere of plasma, instead of a sphere of atomic gases. In a plasma sphere the mass density is primarily determined by the electric-force interaction of the plasma particles, and only secondarily, and rather slightly, by the gravitational force. In this case, for the plasma Sun platform, the recognized mass-density is just about what one would expect, which actually proves the plasma Sun platform. Nor would a gas-sphere with a 1000 times the volume of Jupiter actually be able to exist, as the gigantic gravitational forces in such a sphere would crush all atomic structures within it, whereby the hydrogen Sun theory is self-rendered an impossible dream.

Solar activity
and sunspots

NASA - TRACE image 17.1 nm

Everything that we observe about the Sun comes to light as a
paradox under the hydrogen Sun model, which the plasma Sun
model resolves. The features that you see here, including the
sunspot, are clearly features of electrodynamics, instead of being
features of gas dynamics.

A simple comparison

A simple comparison makes the fact rather plain that the Sun is a vast scene of electrodynamics of numerous types.

Not scenes of gas movements

These are not scenes of gas movements, but are electric phenomena that can only be observed in specific light bands that correspond with electric agitation of atomic elements in which the electric features are revealed.

Sunspots are electric phenomena,

NASA illustration

Even the sunspots are electric phenomena, rather than gas phenomena, as this NASA art illustrates.

Sunspots are holes in the photosphere

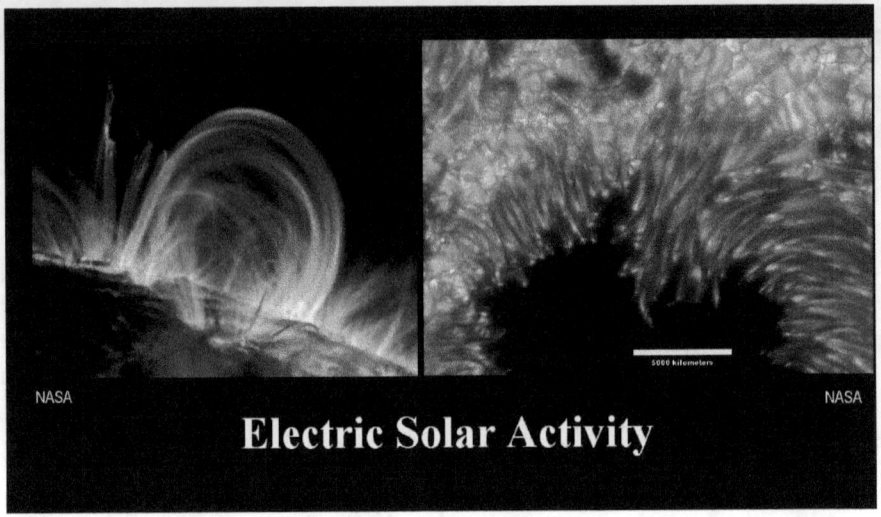

Electric Solar Activity

The sunspots are holes left behind in the photosphere after large-scale electric overload eruptions, and they reveal the Sun to be dark on the inside. Is this what one would expect to see on a hydrogen Sun that is heated from within? Of course not.

Solar wind flowing against the force of gravity

http://www.zam.fme.vutbr.cz/~druck/Eclipse/ - an example of the amazing solar eclipse photography of Miloslav Druckmueller

Nor would one expect to see vast streams of solar wind to be flowing away from the Sun forever and ever, so it seems, against the force of gravity. Neither should these streams of material be accelerating as they flow away from the Sun, to such high speeds as 800 km/sec, if they were gas phenomena originating from a hydrogen Sun. Of course what you see is what one would expect for a plasma Sun, where streams of plasma particles are electrically self-accelerating by the electric force that causes a plasma's self-expansion, in which atomic elements get to ride along.

Corona several million degrees

by Luc Viatour / www.Lucnix.be

Of course, if the Sun was a hydrogen Sun heated from within, it would be impossible for its surrounding atmosphere, its corona, to be several-hundred times hotter than the Sun is itself. Atomic temperatures up to several million degrees have been observed in the solar corona, above a Sun that is itself only 5500 degrees hot. What you see here is natural only for a plasma Sun, resulting from accelerated plasma interactions. Nor do the hydrogen Sun paradoxes end at the Sun itself.

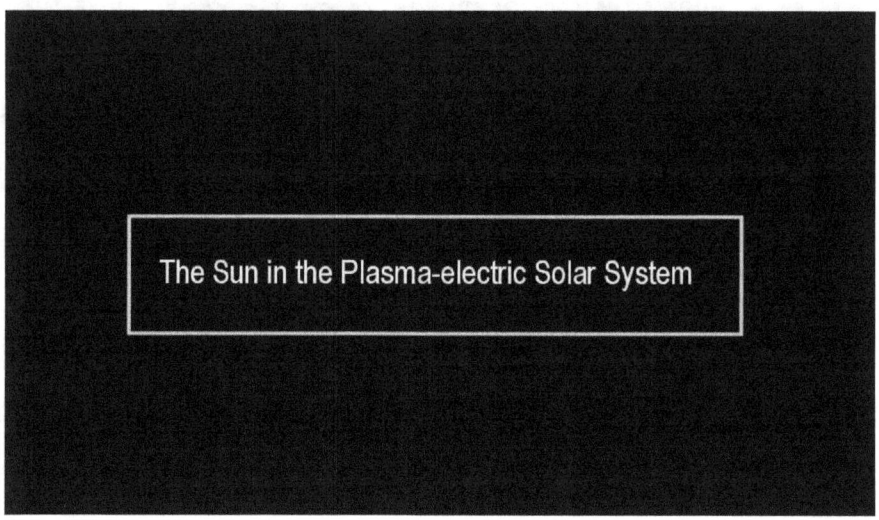

** The Sun in the Plasma-electric Solar System

The entire solar system is electromagnetically organized

The fact that the planets of the solar system orbit in an ecliptic plain, geometrically spaced in distance, in near circular orbits, with their orbits being maintained for billions of years, regardless of cosmic drag, proves that the entire solar system itself is electromagnetically organized. Nothing what you see here can be attributed to the platform of an electrically dead solar system with an electrically neutral hydrogen Sun.

The Sun as a plasma-powered Sun

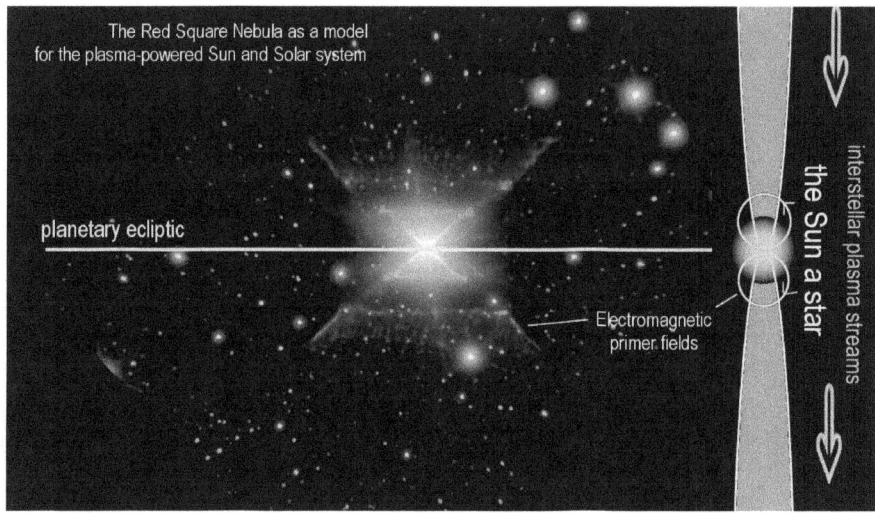

The Red Square Nebula as a model for the plasma-powered Sun and Solar system

planetary ecliptic

Electromagnetic primer fields

the Sun a star

interstellar plasma streams

Only when one begins to see the Sun as a plasma-powered Sun, that itself cannot exist without the vast electrodynamic forces of interstellar plasma streams that power it...

Climate reflects solar cosmic-ray fluctuations

can one acknowledge that nothing is paradoxical from the largest features in the solar system, to the smallest features of the climate on Earth that reflects solar cosmic-ray fluctuations that are inherent in the electric resonances of the solar system as a system of interacting electrodynamics.

Paradoxes with the hydrogen Sun model

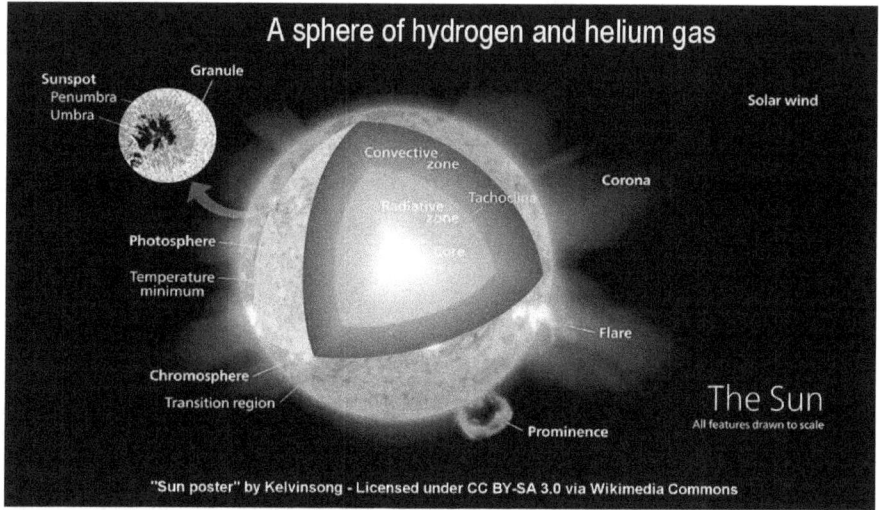

The vast array of paradoxes that we encounter everywhere with the hydrogen Sun model, from the large planetary scale all the way down to that we find reflected in the sunlight that we see everyday - which the plasma Sun platform resolves in every case - leaves one to ask why the hydrogen Sun model remains the most widely accepted model in astrophysics to the present day, in spite of the vast array of self-evident paradoxes that disprove the false, terrible model in every case.

*Terror Game with a False Model

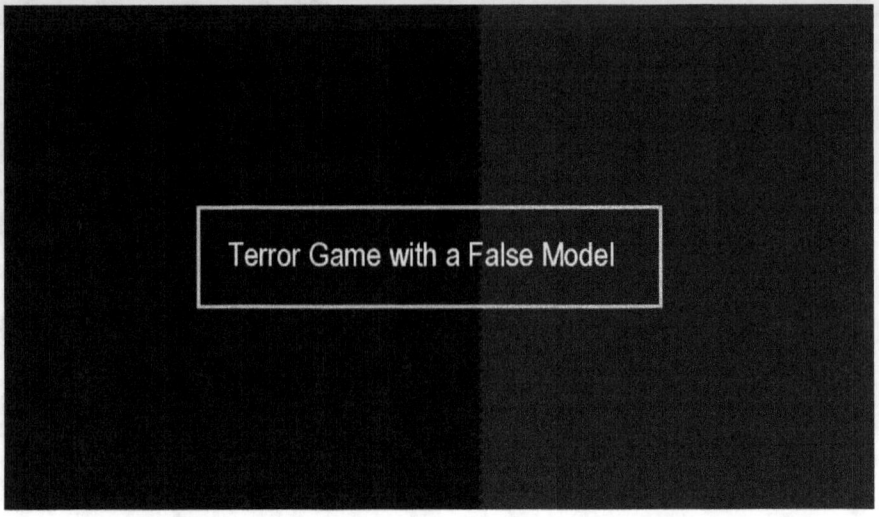

** Terror Game with a False Model

Answers found in the political background

Lord Mayor of London - John Stuttard - Nov 2006 - by Diliff - Own work. Licensed under CC BY 2.5 via Commons -

The needed answers may be found in exploring the political background from which the terribly incorrect hydrogen Sun model is promoted, and what the objectives of the terror are that are pursued by the promoters of this model of the Sun that is presently used to strangle humanity, to cause economic suicide, mass genocide, even depopulation.

Not surprisingly, the path leads directly to the gilded halls of empire, to the 'system' of empire that intentionally employs the false model of the Sun as a foundation to enforce its coveted goal for world domination.

The thoroughly false model

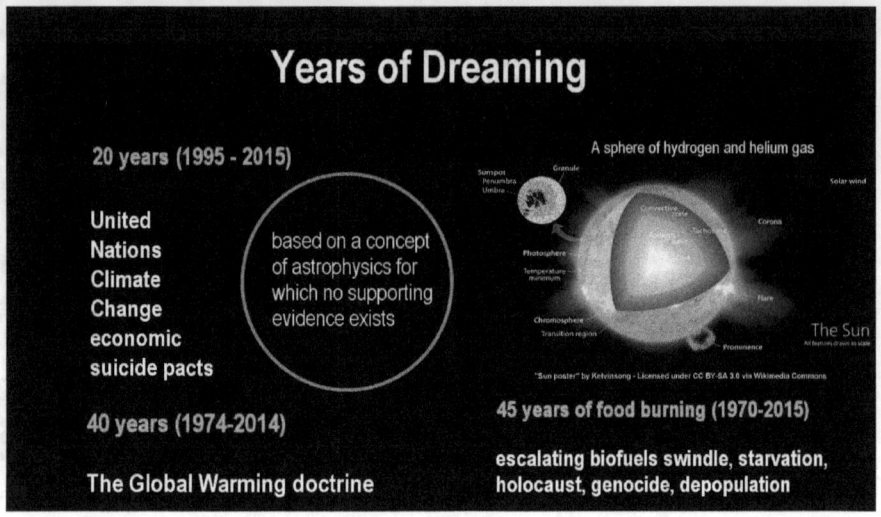

The thoroughly false model is a perfect platform in the hands of empire for inflicting global economic suicide on demand, towards the mass-depopulation of the world by starvation, which the global-warming scare hoopla - for which no real supporting physical evidence exists - is intended for.

The promoters of the counter-science hoopla

Herbert George Wells in 1943
the man who changed
the course of science

wikipedia

The promoters of the counter-science hoopla that stands behind the crippling scene of science devolution, say to society in essence, "The truth be damned! Close your mind. We tell you what the truth is. We own your science. We control what science is. We use this control to prove to you what we want you to believe. Your place is to simply believe. We even supply your opinions for you, that we want you to imagine are real. We own you, by imprisoning you into dreams. You have no means to escape from this trap, because you lack the scientific power to discern what the actual truth - based on real science and real evidence - is. We dazzle you with sparkling dreams, so that you will never reach out for your potential in real science and develop your power as human beings."

H. G. Wells was one of the early pioneers

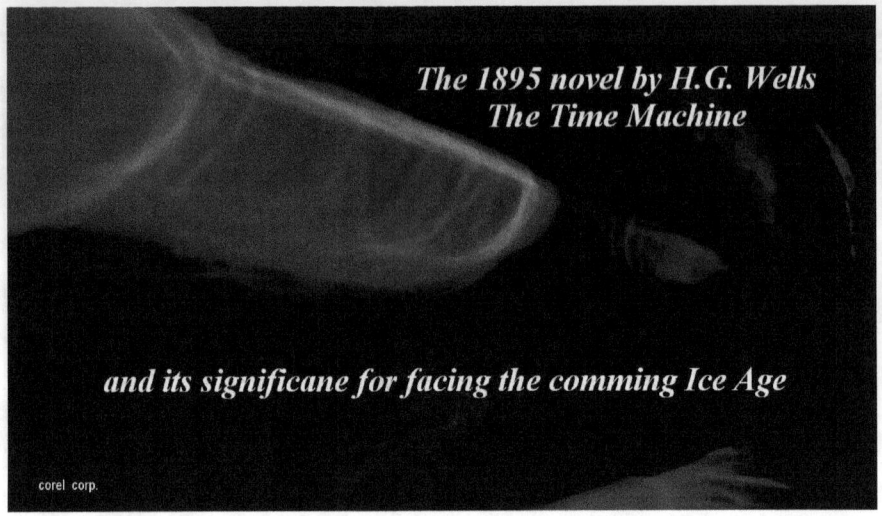

The 1895 novel by H.G. Wells
The Time Machine

and its significane for facing the comming Ice Age

corel corp.

These are the types of lyrics in the songs of empire, which are rarely ever sung openly, though they are never really hidden either. H. G. Wells was one of the early pioneers who had mastered the art of inspiring the terrifying lyrics that started the war against science within the empire, which by now most likely has far exceed Wells' expectations all the way to hiding the science of the Ice Age dynamics whereby the needed preparations will not be made for the coming transition to it in the 2050s, potentially.

Example (image): H. G. Well's novel, "The Time Machine" A story in the distant future where the Eloy (the nobility that has lost its humanity) are kept as life stock for use by the Morlochs (the dirty machine people, the result of scientific and technological progress.) The message to the oligarch is: You must either scrap science (the enemy of oligarchy) or keep it tightly controlled and deployed for your purposes.

The devolution of truth in science

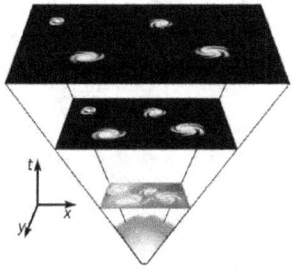

The Big Bang creation myth refuted
by the electric solar fusion model

The devolution of truth in science - with the tool of irrationality - is intentional. The process has been long in progress. The masters of empire have openly admitted that the subjugation of science under their control, is their intention, in whatever form this can be achieved, and against whatever truth may oppose them.
Image: The Big Bang Creation myth.

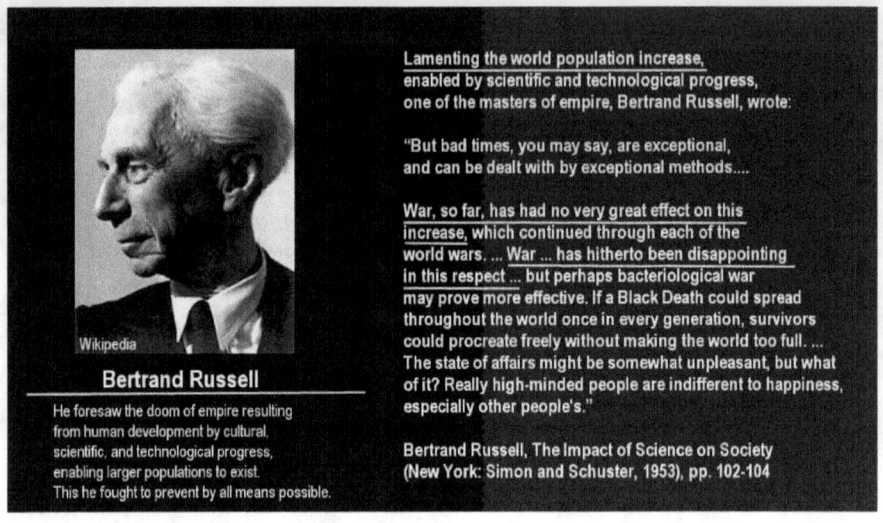

Lamenting the world population increase,
enabled by scientific and technological progress,
one of the masters of empire, Bertrand Russell, wrote:

"But bad times, you may say, are exceptional,
and can be dealt with by exceptional methods....

War, so far, has had no very great effect on this
increase, which continued through each of the
world wars. ... War ... has hitherto been disappointing
in this respect ... but perhaps bacteriological war
may prove more effective. If a Black Death could spread
throughout the world once in every generation, survivors
could procreate freely without making the world too full. ...
The state of affairs might be somewhat unpleasant, but what
of it? Really high-minded people are indifferent to happiness,
especially other people's."

Bertrand Russell, The Impact of Science on Society
(New York: Simon and Schuster, 1953), pp. 102-104

Bertrand Russell

Wikipedia

He foresaw the doom of empire resulting
from human development by cultural,
scientific, and technological progress,
enabling larger populations to exist.
This he fought to prevent by all means possible.

Lord Russell argued that control of science is not enough. He argued in his book, "The Impact of Science on Society," that radical depopulation is the only sure path for the stable existence of empires.

His radical stand is probably rooted in the great debates over 'what to do with science,' that started with Well's call to attention, to mobilize an anti-science direction within the system of empire. Within the debates, the advance of science became evermore recognized by the masters as a mortal threat to every form of empire - a threat that would endanger the very foundation of its system with the unbridled advance of scientific and technological progress.

The debate was centered on whether science should be crushed, or whether it should be reformed to operate as a tool for the purposes of empire, under strictly imposed controls. It appears that the latter path has been chosen.

Contradictions that become mind-killers

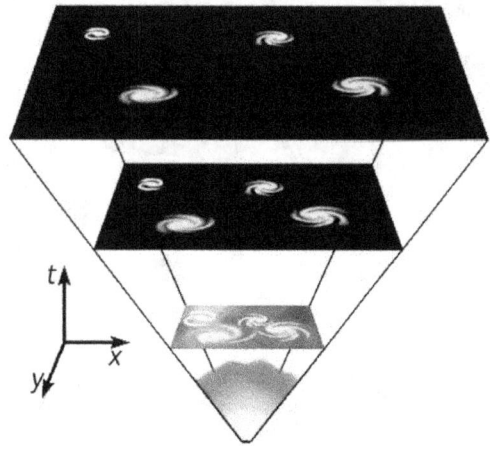

Science became smothered with irrational concepts and contradictions that become mind-killers, such as the big-bang theory that promotes universal entropy, wrapped in theories of star explosions, black holes, and the devolution of everything towards the death of everything, even the death of stars and the death of civilizations, along with the death of sovereignty that is the heart of culture, leading to the death of nations, etc..

The Big Bang travesty

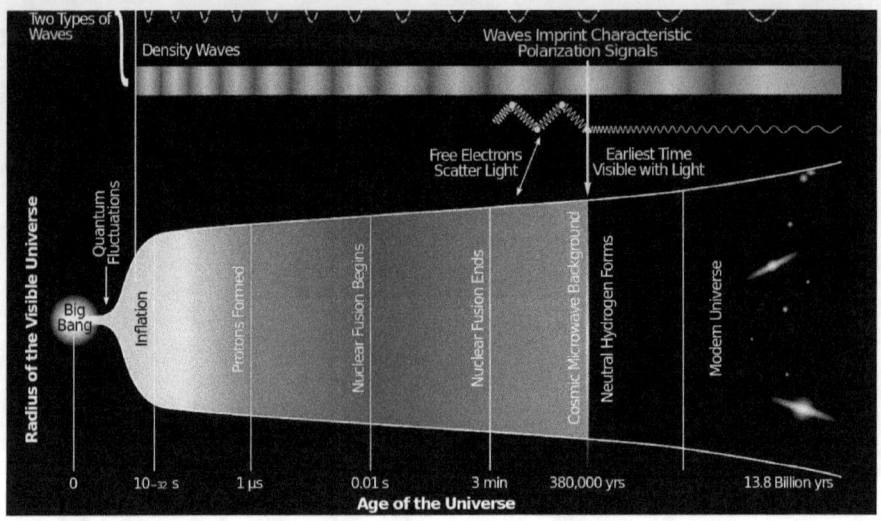

The Big Bang travesty, as concept of a universe that started with a bang and is inevitably fizzing into nothing thereafter, has become a part of the empire song that desires devolution. The devolution concept is blending with the lyrics for the mass-depopulation of the world to less than one billion people, which is now demanded as a means for restoring the environment of impotence that imperial systems require to exist.

For the intentional devolution of science

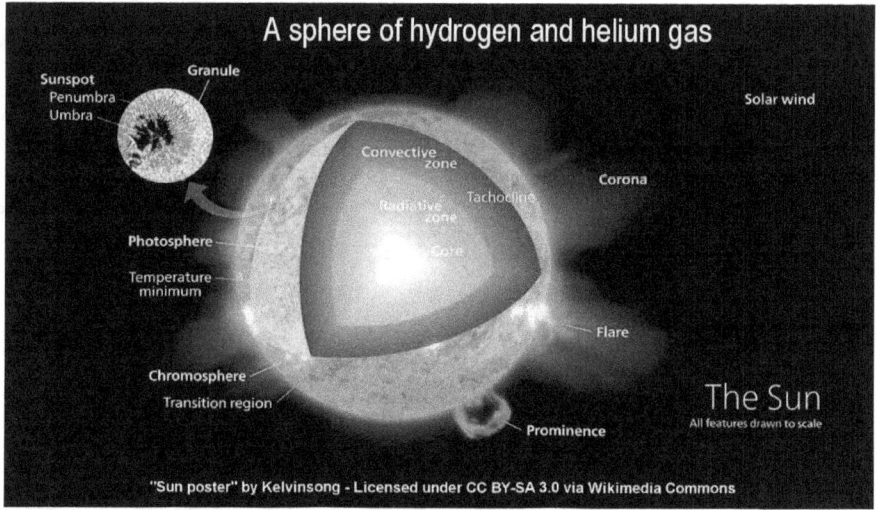

The hydrogen Sun model is one of the elements for this chain created by the masters for the intentional devolution of science and for the controlled distortion of reality that enables the controllers to assert power over society. The false model was developed in the general timeframe of the great debate in the halls of empire over 'what to do with science.'

*Pre-existing Science Inverted

** Pre-existing Science Inverted

The Alfven model of the galactic system

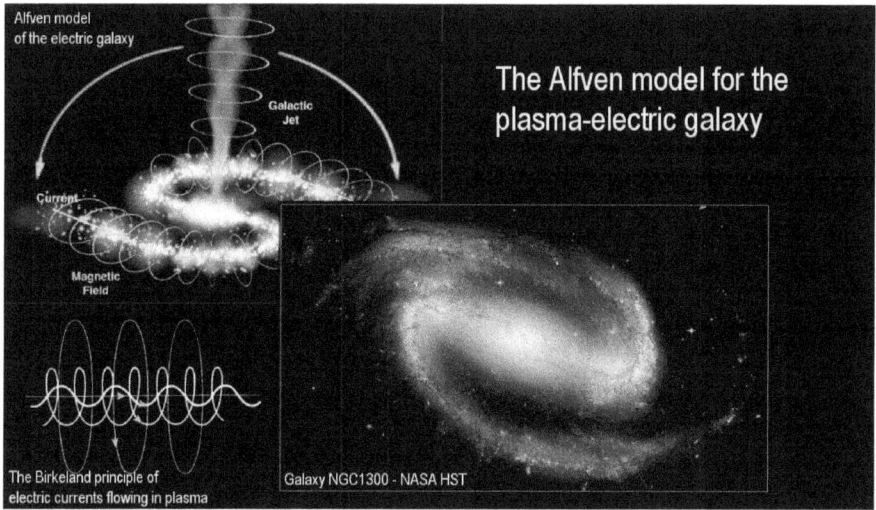

The plasma universe concept is not something new. It had actually
existed prior to the big-bang promotion that the hydrogen Sun
model is a part of.

The plasma universe concept was developed in the early 1900s by
the physicist, and later Nobel laureate, Hannes Alfven. He had
developed a model for a plasma-powered galaxy based on the
leading-edge principles of electrodynamics that became known at
the time, such as the principle of the Birkeland currents that reflect
the discovered characteristics of magnetically self-aligned electric
currents in plasma, discovered by the Norwegian explorer and
physicist Kristian Birkeland.

The Alfven model of the galactic system, as a system of self-twisting
Birkeland currents, had established a revolutionary platform for
perceiving a galaxy in terms of an electrodynamic galactic system
that accords not only with known physical principles, but also
accords to the present day with the observed motions of stars in
the real galaxy.

The truth be damned!
All stellar motions
MUST
be seen as orbital motions

NASA - galaxy M51

Alfven's truthful model became quickly overlaid with a model that stands in complete opposition to all known principles in astronomy. In the resulting counter-science project, a galactic model was imagined that had all the stars in a galaxy orbiting an imaginary galactic center of a super-massive black hole. This became the widely accepted model, promoted throughout the sciences. Here the built-in paradox, which was evidently intentionally created, is that the model is physically not possible according to all known laws of orbital dynamics. An orbital system on the scale of a galaxy is simply not physically possible.

The goal to crush the heritage of Johannes Kepler

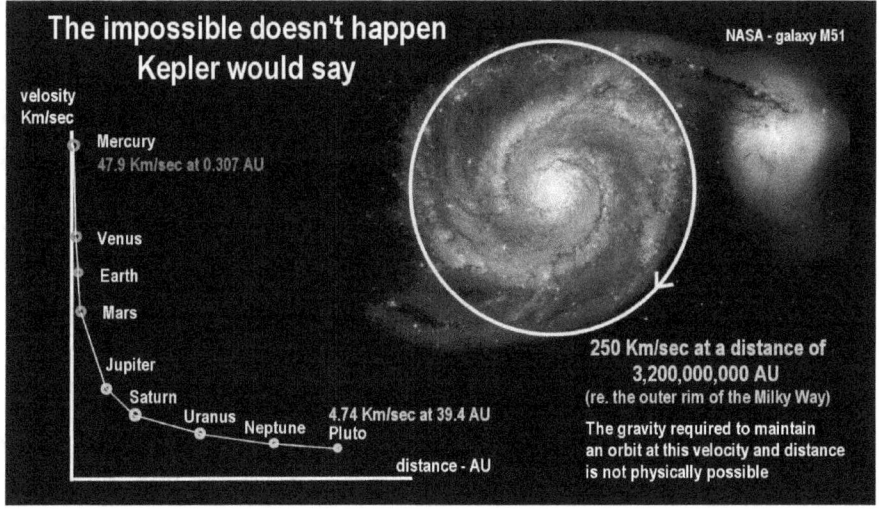

If the goal of the false model was intended to crush the heritage of the great pioneer in astronomy, Johannes Kepler - one of the great pioneers of science - and to crush science itself as a concept related to truth, the project to reach this goal has succeeded. The laws of orbital motions that Kepler has discovered, being reflected in the solar system, became applied to the galactic scale in total disregard of the physical principles that underlie orbital motions, which render orbital motions impossible on the gigantic galactic scale due to the rapidly diminishing reach of gravity.

Orbiting stars totally impossible

Stellar orbits arround the galactic center are NOT possible under Keper's laws of gravitational mechanics. However, no other perception is allowed. Thus new mythical 'epicycles' are imagined to make the modern doctrine plausible.

When it became self-evident that the theory of orbiting stars is totally impossible under the Kepler-discovered laws of orbital motion, society quickly became smothered with evermore irrational concepts to maintain the false theory, which were intended to render the impossible, plausible. For this, the almost desperate cover-up project of the concepts of dark matter and dark energy were imagined. These are phenomena that no one can actually see or detect with technology, but which are required for the fraud to function, which serve as a type of epicycle to twist the evidence to match the impossible.

** Beyond World Conspiracies an Open Door to Live Again

Lord Christopher Walter Monckton in Washington, D.C.
Head of the Policy Unit United Kingdom Independence Party
Photo © 2009 Joanne Nova (CC-BY-SA) Wikipedia

COP15
COPENHAGEN
UN CLIMATE CHANGE CONFERENCE 2009

Global Warming Fraud exposed by Viscount Lord Christopher Monckton

Image: Logo of the 2009 U. N. Climate Conference in Copenhagen. It appears that the various efforts in science dissolution have succeeded. More and more concepts that are physically impossible are now fully accepted and promoted by all western scientific institutions. Society is told by the empire-subjected institutions and world institutions, "see, what is impossible is really happening; and see what is scientifically false, is in the New Age science universally regarded as true. And, in case you are still confused, the consensus of scientists is that the CO_2 molecule in the air that you contribute by living and breathing, is a dangerous greenhouse gas that will overheat the Earth and eventually make life impossible, so please, do commit economic suicide by shutting down energy production, and with it your living, and please do commit mass-genocide by the massive burning of food in automobiles to save the planet. You are even pre-absolved by us, of the crime of committing murder."
This is how society becomes coerced to lay itself down to die at the end of the long road of intentional science devolution.

The sunlight to unravel the paradoxes

It is evidently hard for society to dig itself out of the deep grave of science devolution that has been decades in the making, to bury it. Nevertheless, the sunlight that we see every day is white, and it has absorption lines spangled across it, which are features that only the plasma Sun can facilitate, which also facilitates solar cosmic-ray flux in changing amounts that control the climate on Earth. Society may begin here, with the sunlight that is seen everyday, to unravel the paradoxes.

If one needs proof that our Sun really is a plasma Sun, the most-visible proof is located right here in the sunlight. That's a good point to start with, because no other model for the Sun, than the plasma Sun model, is capable of producing the phenomena that you see before you.

Humanity has a model it can trust to be real

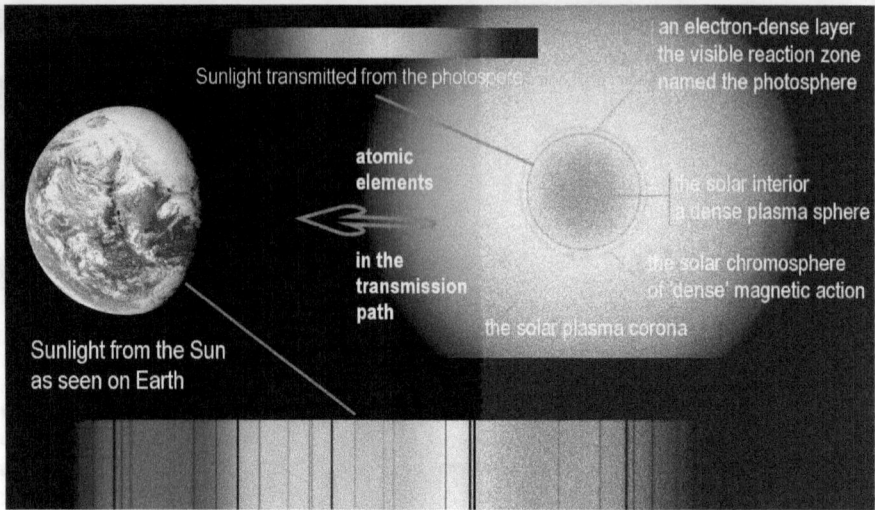

This means that humanity has a model before it that it can trust to be real. This also means that the known climate consequences that correspond with this model, can be trusted likewise, and that the consequences based on contrary models can be safely scrapped.

The entire pyramid of tragedies relegated to the trash bin

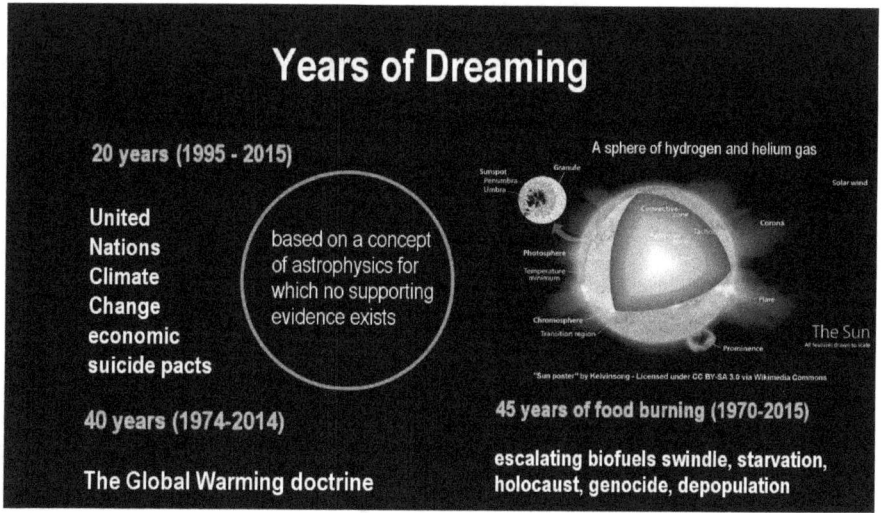

This means in real terms, that the entire pyramid of tragedies that are rooted in the long-paraded false model of the Sun, from global warming, to greenhouse scares, to biofuels holocausts, and all the way to depopulation policies, can be relegated to the trash bin, which sets us free to move forward with what we know to be real. With the vast scene of inverted science and its consequences put behind us, and with every vestige of the tragedies built on this trash fully purged from the face of civilization, we become free at last, to live again.

With the truthful model of the Sun established

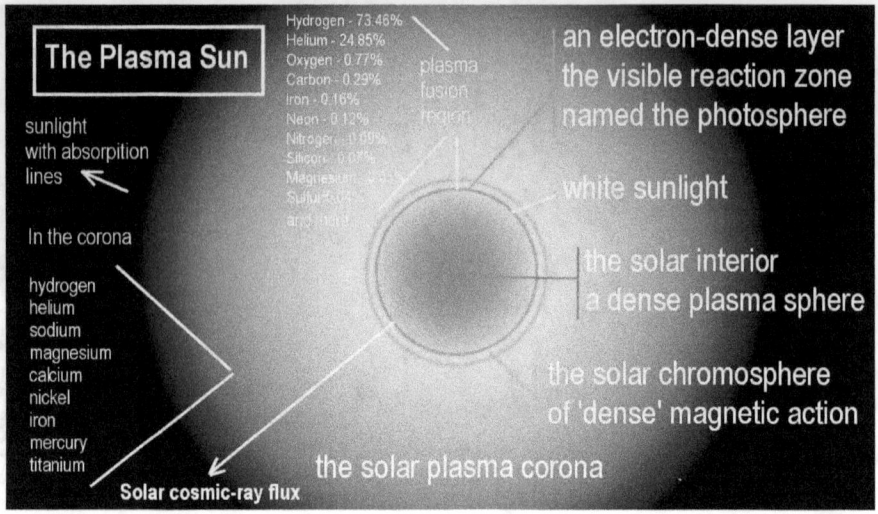

With the truthful model of the Sun established, based on real physical evidence, the years of dreaming that have shackled humanity for half a century with illusions, end. With truthful science established, including the science of the Ice Age dynamics, the science doldrums come to an end that have allowed centuries of society lying to itself, such as in the form of terrorism, war, and the preparations for nuclear war. Even economic looting and economic collapse, will end when science inspires discoveries of truth, and the understanding and acknowledgement thereof. On this path with wider and truer horizons the impending new Ice Age comes to light with greater challenges for humanity to meet, than have ever been faced before in all history, but which hold the potential for us, in meeting the challenges, for us to become more intensively human than society allows itself to be in the present.

Inexhaustible energy resources for an infinite future

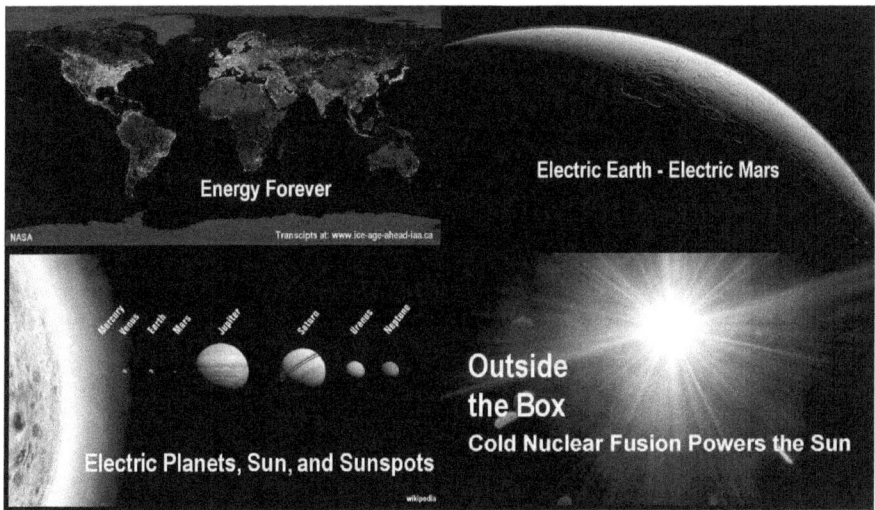

The plasma Sun model also offers inexhaustible energy resources to mankind for an infinite future, since the cosmic plasma streams that power the Sun are also available to be tapped into by humanity with technology. The age if inexhaustible energy begins here.

A basis for understanding the ice ages

The plasma Sun model even provides us a basis for understanding the dynamics of the ice ages with verifiable, real physical evidence and the science to forecast the future with a high degree of accuracy, and with enough time to spare for the needed preparations for the next Ice Age that all evidence tells us will likely begin in the 2050s, with the plasma Sun going inactive altogether at this time. Without the plasma Sun model being recognized, and it being understood and acknowledged, the science platform in the world would remain so small that humanity won't meet the Ice Age challenge, or even recognize it, and become overwhelmed by it and die.

A new era of renaissance stands before

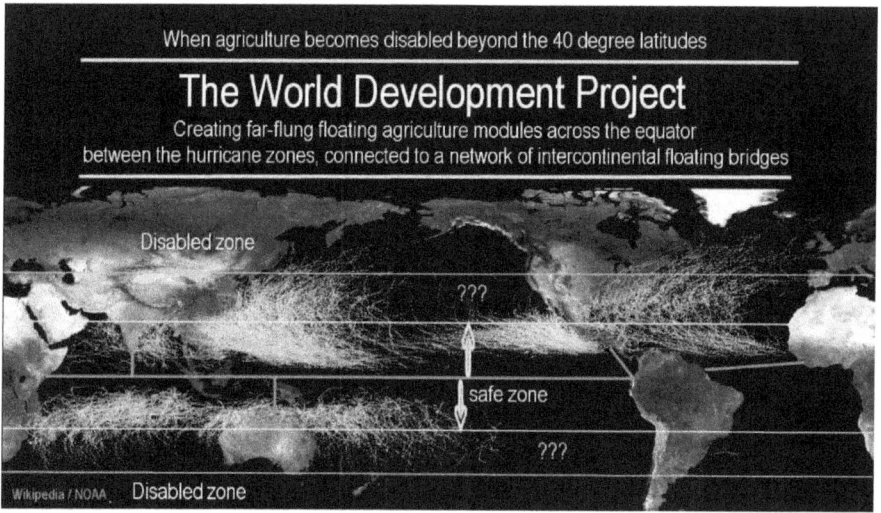

But with the science breakthrough established, a new era of renaissance stands before us in which not even the sky is the limit, where anything is possible. Then, we will build the needed infrastructures and meet the coming Ice Age with a song.